D1557292

DRYING UP

UNIVERSITY PRESS OF FLORIDA

Florida A&M University, Tallahassee
Florida Atlantic University, Boca Raton
Florida Gulf Coast University, Ft. Myers
Florida International University, Miami
Florida State University, Tallahassee
New College of Florida, Sarasota
University of Central Florida, Orlando
University of Florida, Gainesville
University of North Florida, Jacksonville
University of South Florida, Tampa
University of West Florida, Pensacola

DRYING UP

The Fresh Water Crisis in Florida

JOHN M. DUNN

University Press of Florida
Gainesville · Tallahassee · Tampa · Boca Raton
Pensacola · Orlando · Miami · Jacksonville · Ft. Myers · Sarasota

Printed in the United States of America on acid-free paper

This book may be available in an electronic edition.

24 23 22 21 20 19 6 5 4 3 2 1

Library of Congress Control Number: 2018943428
ISBN 978-0-8130-5620-3

The University Press of Florida is the scholarly publishing agency for the State University System of Florida, comprising Florida A&M University, Florida Atlantic University, Florida Gulf Coast University, Florida International University, Florida State University, New College of Florida, University of Central Florida, University of Florida, University of North Florida, University of South Florida, and University of West Florida.

University Press of Florida
2046 NE Waldo Road
Suite 2100
Gainesville, FL 32609
http://upress.ufl.edu

Contents

Introduction

ON THE MORNING OF JUNE 18, 2012, the residents of Cedar Key, a picturesque, fishing-island community on Florida's gulf coast, awoke to what they thought would be another normal day. They turned on their faucets to get water to make coffee, or to shower, shave, or brush their teeth. The water tasted normal.

But there was a problem—a big one.

As the locals and tourists went about their morning rituals, the town's water utility staff members were taking periodic monitor readings of the community water supply and were shocked. They'd been tracking a noticeable uptick of salt in the water for several days, but now "the readings for chloride were so high, and the change happened so fast, that the men thought the testing equipment was broken and sent it off for repair," recalls John McPherson, general manager for Cedar Key Water and Sewer District.

The people of Cedar Key are used to a little salty flavor in their water. After all, they like to kid themselves by saying that whenever they go out of town and order a coffee, they always add a little salt to make it taste right.

This time, though, salinity in the water was no joke. The next day everyone in town learned that Florida's Department of Environmental Protection (DEP) was shutting down their water supply because the salt level was dangerous to human health. There was nothing wrong with the water testing equipment.

Later that day donations of bottled water from the nearby towns of Bronson and Chiefland arrived on trucks to keep everyone hydrated until officials could figure out what to do.

What had gone wrong? "It was a combination of drought and pumping too much," says McPherson.

And maybe sea level rise.

City officials next installed a new reverse osmosis system that converts salt water to fresh when needed; not long afterward two tropical storms blew in and recharged the aquifer, pushing the salt water back down into the ground underneath a layer of fresh water. The town's problems weren't over yet, however. For a while, the fresh water filled with iron and organic compounds that kept clogging up the filters of the new system.

Then in September 2016, Hurricane Hermine blasted into town and caused more grief for the locals of Cedar Key. Flooding ruined homes and businesses. City waterlines broke and drained the town's landmark freshwater tower. Stormwater lines filled with seawater that disrupted the treatment process of stormwater at a nearby facility.

Today Cedar Key's water supply problem has been fixed, but perhaps just for the moment. "There's a sense in town that Cedar Key is imperiled," says McPherson. "It is so low lying, that before our well fields are destroyed, our town will be destroyed. It's already flooding with seawater."

By now, he adds, you start to wonder, "At what point do you stop investing in the city's water supply and give up because it's futile?"

It's a sobering, yet logical, question. After all, without water there's no point of having a community.

What Cedar Key's residents have already endured should serve as an early warning system for much of the state. They, after all, aren't the only Floridians contending with water woes. In fact, you have to search hard to find a Florida community that doesn't have a water problem, or soon will have.

Sea rise and drought aren't the only culprits. In 2007 Cynthia Barnett, in her award-winning book *Mirage: Florida and the Vanishing Water of the Eastern U.S.*, warned Floridians their drinking water was imperiled.

A main reason? Overconsumption. "Today, Floridians are pumping groundwater out of their aquifers faster than the state's copious rainfall can refill them," Barnett wrote. "Meanwhile each new master-planned community, shopping mall, and highway drains water in a bit of a different direction and lowers groundwater levels a little bit more."

In an interview with National Public Radio's Renee Montaigne, Barnett said, "Florida's groundwater has been over allocated—not just in South Florida, but also throughout the state. In addition, we just haven't taken conservation as seriously as other parts of the country. People are in denial."

In that same year, John Mulliken, director of water supply for the South Florida Water Management District, told the Associated Press, "We just passed a crossroads. The chief water sources are basically gone. We really are at a critical moment in Florida history."

Barnett's and Mulliken's analyses still ring loud and true. Today, Florida's water problems loom large and are likely to worsen, especially if Floridians continue to treat water as a business-as-usual matter.

In fact, many Florida researchers believe that if their state remains on its current population growth and development trajectory, eventually there probably won't be enough good, clean water for everyone—humans, wildlife, and ecosystems alike. In short, what Floridians are doing is clearly not sustainable.

At the same time, the state must also contend with periodic flooding episodes that are more likely to worsen thanks to both nonstop urbanization and powerful storms. This situation means billions of gallons of water will keep draining out to sea. Only in Florida, perhaps, would you hear water managers warning of coming shortages as they simultaneously plan to control flooding by piping "excess" fresh water so deep into the ground that it can never be retrieved.

Meanwhile, there's a growing welter of water pollution problems from Key West to the Georgia state line. Too many of Florida's rivers, lakes, springs, marshes, wetlands, estuaries, and even aquifers—or underground water reservoirs—are now menaced by a slew of dangerous toxins, ranging from algae to chemical and radioactive substances.

Making matters worse, huge earth-moving machines relentlessly destroy Florida's farmlands, forests, marshes, and wetlands to make way for new strip malls, office spaces, schools, gas stations, and communities, some the size of cities. In so doing, they are not only destroying habitat, but also disfiguring the state's hydrology—that is, the state's natural plumbing and water cycle.

Floridians, of course, are just a small fraction of the seven billion human beings on the planet who find themselves on the threshold of a new epoch—one defined by runaway technology, global trade, and transportation, along with the rapacious water demands of a growing human population and the unfolding impact of climate change.

But what is so jaw-dropping about Florida is that severe water shortages are emerging in a state that ought to be otherwise saturated. All Floridians are never far from a coast, river, or lake. Wetlands, marshes, swamps, bogs, water retention ponds, canals, and a galaxy of swimming pools are everywhere. The very ground underneath your feet may have been wet and mushy, even underwater, not too long ago—that is, before a developer filled it in with sand and muck from a nearby river or wetland. In fact, if not for a breathtakingly bold, vast, effective, and destructive engineering assault on the state to drain its water, much of the Sunshine State would still be water soaked and uninhabitable. Even if you live in Florida during a drought, which hits the state on average at least once every ten years or so, you will still be in a wetter place than most of the nation's arid southwest.

Rain falls a lot in Florida, averaging 52 inches every year. A lot of it accompanies hurricanes that appear in late summer and early fall, soaking Florida's flattish landscape, recharging—sometimes overcharging—the natural water supply. Florida receives more of these destructive storms than any other state.

Even the air is thick with water. In fact, Florida is the most humid state in the country. With the seas creeping inland and up from below, and an ever-warming climate, you can expect the future to be even more humid.

The state is dotted by more than seven thousand freshwater lakes, including the biggest lake in the South—Lake Okeechobee. More than seventeen hundred streams and rivers flow across the state, the largest of which is the troubled, lumbering, polluted, diverted Everglades.

North Florida boasts the world's greatest cache of freshwater springs, more than nine hundred at last count. Its estuaries—those coastal areas where fresh water and salt water merge to provide aquatic habitat—are among the most productive in North America.

Water is vital to Florida's economy: tourism, agriculture, fishing, industry, military, government, aerospace, transportation, commercial, and other sectors all rely on it.

Florida's flora, fauna, and fragile ecosystem depend on a rich supply of fresh, clean water.

Water is also central to a true Floridian's identity. Millions of residents and tourists alike recreate in the state by fishing, swimming, surfing, kayaking, skiing, and taking part in a host of other water-based pastimes.

Today, however, Florida water is under siege. And it's no secret. Travel across the state, and you'll discover a growing array of water seminars, conferences, and forums attended by legions of water managers, consultants, government officials, scientists, researchers, engineers, political figures, environmental activists, urban planners, journalists, agriculturalists, business leaders, and worried citizens who are well aware that their state has persistent, worsening, and growing problems.

But they don't agree on the causes, nor the solutions. For instance, many of Florida's stakeholders prefer to stick with the status quo—that is, altering natural systems and landscapes to accommodate the increasing water needs of humans. Others prefer a much newer and radically "bluer" approach.

But will any of these approaches work as long as Florida remains a state whose powerful leaders continue to welcome runaway population growth, development, and sprawl even at the expense of the environment?

Water disputes have already occurred in Florida, and new ones keep popping up all the time. During the past half century hundreds of lawsuits have been filed over issues related to consumptive water permits, water pollution, wetlands protection, and stormwater and wastewater management. There are also never-ending disputes over water usage, storage, restoration, distribution, and pricing.

Some of these quarrels have evolved into so-called water wars at both the regional and interstate levels. Meanwhile, water battles elsewhere in the world have prompted military and security experts to warn of a global disorder if water isn't fairly allocated among all parties.

Once you delve into Florida's water issues, you quickly find that there's much to think about. For example: How did these prickly issues come about? How should they be resolved? Who is entitled to water?

How much should they pay? Who decides how much anybody gets? What should it cost? Should private companies or the government be in charge of water resources? What are the best ways to protect, conserve, and use water?

The overriding goal of this book is to shed light on many of the biggest and most vexing of Florida's mounting water problems and what is being done—and not being done—about them. The author's hope is that such information will help readers become water-literate enough to take part in the discussions and actions that will determine the fate of their share of the most precious substance on earth. It's much too important an issue to be left solely in the hands of politicians, the business community, developers, farmers, and engineers.

It's true that Florida's water supplies may not be literally "drying up." But they are being squandered and polluted to such a high degree that there may not be enough to go around to satisfy the needs of all.

It's not too soon to begin to act. Some water experts, in fact, argue that Florida is fast approaching a tipping point concerning the well-being of its water resources. Others think it's too late, and the state is already in big trouble. A few, however, insist there's endless water in Florida's aquifers, and there's no emergency. Denial, however, invites disaster. Just because water has always been there doesn't mean it always will be. Just ask the people of Cedar Key.

1

Business as Usual?

IT WAS TIME FOR CHARLES DRAKE TO SPEAK. Polite applause greeted the confident-looking, middle-aged man as he took the podium at the 2016 American Industries Water Forum in Orlando. "You shouldn't clap until you hear what I have to say," Drake said, deadpan. Then he clicked on his PowerPoint show.

The reason for his air of gloom became clear when he introduced his topic: "Are We Ready for Direct Potable Re-use Water?" (That is, are Floridians prepared to drink treated wastewater?)

Drake, a member of the St. Johns River Water Management District (WMD) and an Orlando-based executive with Tetra Tech, a global water engineering company, explained that recycling water is a smart way to respond to Florida's unfolding water shortage. After all, the state already leads the nation in using treated wastewater to irrigate farm fields. And now the time had come, he told his audience, to get serious about turning wastewater into drinkable reuse water. It is a cheaper and easier way to produce a fresh water supply alternative than other contenders, such as the desalination of seawater.

There are two ways to get potable reuse water into a drinking glass on the kitchen table. One is for the nasty water to get scrubbed up in an advanced water treatment plant and then be allowed to seep back into the ground through layers of earth for extra cleansing before it mixes with the underground water supply. The other way—Drake's main point—is to use new, more advanced wastewater treatment technology that can render wastewater that's pure enough to be pumped directly to a city utility system for you to drink.

To lighten things up a little, Drake clicked on a video clip from Jessica Yu's documentary, *Last Call at the Oasis*—a spoof of his own talk—that featured comedian Jack Black, playing himself preparing to perform in an infomercial touting a new kind of bottled water.

Before the shoot begins, a clueless Black asks for clarification about the recycled water product he'd agreed to promote. His business manager, trying to be reassuring, explains it has to do with this "whole, crazy water thing going around the world." Then he gets to the point. "They're taking water from gutters . . . sewers . . . your toilet . . . and then there's some sort of . . . scientific process, I don't know, above my head, right? . . . They're literally straining the poop juice right out of the water that you and I drink. So, it's defecate, drink-acate . . . This is, like, the future."

Black is shocked and disgusted. But after a little coaxing, he mans-up and chugs a whole bottle of reclaimed sewage to a round of applause from the video production crew.

Are Floridians ready to join Black and drink bottled water made from such stuff? Probably not many—just yet. But potable reuse beverages are already being drunk around the world, including in Singapore, where water shortages forced the government to encourage citizens to imbibe lots of what they call NEWater.

Drought-stricken California is experimenting with potable reuse water, but convincing consumers it is safe to drink has been a hard sell.

Clearly, a lot of work is needed to gain public acceptance—that is, to overcome the "yuck" factor—in Florida. The WateReuse Association has already taken steps to help Floridians overcome that hurdle. In July 2016 the utilities trade organization set up various demonstrations in select Florida cities to promote potable reuse water. In Tampa, the trade group offered more than one hundred Florida homebrewers free samples of potable reuse water to take part in a competition to make the best beer.

If all goes well with public education and marketing efforts like the one in Tampa, says Drake, potable reuse drinking water could be available in Florida within ten years.

That water professionals are not only considering, but seriously promoting recycled wastewater—as clean as it may be—as a drinking-water solution underscores how worried, or pragmatic, they are about Florida's water future.

Why Worry?

Water, for the moment, is readily accessible for most Americans. And that might be part of the problem, because we take it for granted. It's just there. Unless you have your own deep well, you probably get your water from a utility that has provided the infrastructure to pump water out of the ground (or from a local surface water body), clean it up so well that you won't die from drinking it, and then deliver it to your house. Turn on the tap. Instant water! Twenty-four hours a day. Every day. Even during the aftermaths of hurricanes, the water is usually available, because utilities have emergency backup systems to keep the water flowing.

Floridians, like most Americans who don't have their own wells, pay for the *delivery* of water, and not very much, if anything, for the water itself.

The days of abundant, cheap water in the Sunshine State, however, may be coming to an end.

How Bad Is It?

Water experts emphasize there's a difference between *water stress* and *water scarcity*. They use an equation to make the point. Water stress happens when the yearly water supplies in a given population drop below 1,700 cubic meters per person. When they sink below 1,000 cubic meters for each individual, water scarcity is in effect. *Absolute* scarcity strikes when the allocation for each person is less than 500 cubic meters.

There are even finer distinctions. According to the United Nations, lands in Sub-Saharan Africa, for instance, suffer "economic water scarcity," a condition that results when a local population is too poor to develop a local source of water. "Physical water scarcity" occurs when there's not enough water to quench demand.

Unlike drought-stricken California, or much of the Mideast at the moment, most of Florida is not enduring economic or physical water scarcity. Children are not dying for lack of water. No young Floridian treks miles on foot to a community well, fills a bucket, and carries it home, as millions of people—mostly young females—do every day in many developing, parched nations.

So the water quantity situation isn't that bleak—yet. But it isn't getting better, either.

Water shortages, in fact, have already arrived in some places of the state. In 2010, according to an independent analysis by the *Atlantic* magazine, the Orlando area ranked tenth among American cities running out of water.

More shortages may be on the way for the rest of the state. The number of locations designated by Florida's five water districts as a water cautionary area is growing. In April 2017 enduring drought triggered a warning of water shortages for 8.1 million people from Orlando to the Keys. (A busy hurricane season that summer ended the shortage for a while.) Looking further into the future, a 2013 study for the U.S. General Accounting Office (GAO) says Florida water managers expect to see regional water shortages by 2023.

For some Florida communities such predictions are too late. "In 1998, Lake Brooklyn was full; a year later the water was gone," says Vivian Katz, president of Save Our Lakes, a grassroots group working to restore lake levels near the scenic small town of Keystone Heights in southwestern Clay County, also known as the Lake Region of North Florida. The community has been hard hit economically ever since its tourist-attracting lakes started to shrink in recent years. Too many sinkholes in the area, not enough rainfall, and sand mining may help explain why they have been disappearing, but Katz and her group believe overpumping of the aquifer by large water users in the Jacksonville area is largely to blame, or at least makes the situation worse.

Katz always wears a red T-shirt emblazoned with the words "Save Our Lakes" when she shows up before the St. Johns River Water Management District governing board to remind them of her town's plight. Feisty and resilient, she's appeared before the board time and again, imploring them to stop the water hemorrhaging in her hometown. "I wish all these people in the audience had on a red shirt, and it scared the Bejesus out of you," she told board members during a public meeting in January 2016. This time Katz has come to object to a rancher's request for a permit to pump millions of gallons of water a day from the already stressed aquifer. "We are the canary in the coal mine. What happened here can happen elsewhere in Florida. We're going to turn into California if we're not careful."

It's not a new issue. "I've been at this twenty-seven years, and there were people before us coming here for twenty-five years."

Florida isn't the only state with these kinds of water supply problems. That same GAO report says that water managers in thirty-nine other states also expect a water shortage somewhere on their home turf in the near future.

Coming water shortages also are expected around the world. "By 2025, 1.8 billion people will be living in countries or regions with absolute water scarcity, and two-thirds of the world's population could be living under water stressed conditions," according to a 2014 United Nations water scarcity study.

There are plenty of reasons for why water supplies dry up or disappear. Sometimes, drought is the devil. The Fertile Crescent in the Mideast, for example, was once a wondrous agricultural belt, until the rains ceased. Today, much of it is desert.

Natural forces and human folly can also team up to create unlivable conditions. Scholars now think the combined impact of climate change, deforestation, and drought explain why the Mayan civilization in Central America, with a population of 19 million, suddenly collapsed sometime between the eighth or ninth centuries AD.

In January 2016 NASA satellite photos revealed that Lake Poopo, Bolivia's second largest lake, had dried up in just three years. Drought and human diversion of water were responsible.

All too often, humans deserve all the blame. Such was the case with the Aral Sea in Central Asia when Soviet engineers in the first half of the twentieth century diverted rivers and depleted the lake's fresh water supply by more than 90 percent, making it one of the worst environmental disasters in history. The Yellow, Nile, and Colorado Rivers also no longer reach their deltas, thanks to human interference.

By June 2017 Bangalore, India's so-called Silicon Valley, was running out of water. The city's water sources were tapped out and polluted. Its 8.5 million residents faced dangerous shortages and depended on private water tankers run by cartels, often called "the water mafia."

The situation was also grave in drought-stricken Cape Town, South Africa's second most populous urban area. In early 2018 officials expected that by July 9 of that same year—a day dubbed Day Zero—Cape Town's 4 million residents would run out of water.

Meanwhile, farmers and ranchers in the Great Plains of the United States are rapidly depleting the Ogallala Aquifer faster than melting snow can recharge it. One of the world's biggest aquifers, the Ogallala lies underneath parts of South Dakota, Nebraska, Wyoming, Colorado, Kansas, Oklahoma, New Mexico, and Texas. If present trends continue, 70 percent of the underground water could vanish in the next fifty years, according to a National Academy of Sciences 2013 report. If that happens, the world's food supply could be in trouble.

Linda Bystrak, a former college teacher and Lake County, Florida, water activist, knows what happens when the water runs out. As a young teenager, she and her family moved to the Bahamian island Eleuthera, where her father worked on a small U.S. Naval base during the Cuban Missile Crisis. Because water was in short supply, the Americans had to both drill a well and transport water to the area. Most of the locals, however, relied on rainwater they caught in cisterns.

Today, the base is gone, and islanders still struggle to secure a stable fresh water supply. "Perhaps if the farms, canneries, and the American bases had not pulled so much fresh water from their aquifers in the '50s and '60s," suggests Bystrak, "they could still meet their need for potable water."

What happened in Eleuthera is a case study of demand exhausting supply. But it doesn't necessarily represent that catastrophe is inevitable. After all, "There is enough freshwater on the planet for seven billion people," says the UN study, "but it is distributed unevenly, and too much of it is wasted, polluted and unsustainably managed."

A report from an international think tank, the World Water Council, is more explicit: "There is a water crisis today. But the crisis is not about having too little water to satisfy our needs. It is a crisis of managing water so badly that billions of people—and the environment—suffer badly."

These patterns can be found in Florida. Billions of gallons of fresh water are flushed out to sea every week, much of it because of human engineering. Fresh drinking water is used to flush toilets and wash cars. Each day, Floridians squander more than 50 percent of the state's daily water usage irrigating their lawns.

Fresh water is also not evenly distributed in the state. South Florida, for example, is densely populated and uses about 50 percent of the fresh water consumed in the state, but most of that water exists in the less-populated north.

Use It and Lose It

When you consider water, it's important to note the difference between water *use*—which simply means the total volume of water extracted—and water *consumption*, which is the amount used, but not returned to its source. For instance, after you wash your car, much of the water you use soaks down into the ground and rejoins the aquifer. So this portion isn't consumed. Some of the water, however, is evaporated or absorbed by fruit on a citrus tree that will be sent to market. A portion of it may dribble down your driveway and end up in a stormwater system where it will be piped away. Add up these amounts, and you have the net loss of water in the area where it was *consumed.*

Both water use and water consumption rates are critically important to the hydrological health of all living things, but too often in Florida, business, political, and economic leaders focus much too narrowly on whether a water supply has enough water for their developments and projects. They often ignore the fact that land use changes disrupt, alter, or even obliterate natural hydrological systems. In fact, these human-caused distortions may be the biggest threat to fresh water supplies.

Wrecking the World's Hydrologic Health

Land use changes are happening at an alarming rate around the globe. The World Wildlife Federation estimates that 31 percent of the planet's land mass is covered by forests, and it reports, "Some 46–58 thousand square miles of forest are lost each year—equivalent to 48 football fields every minute."

There's a cost to this destruction. According to a United Nations report, "Water for Life, 2005–2015," "as human populations grow, industrial and agricultural activities expand, and climate change threatens to cause major alterations to the hydrological cycle."

Some scientists estimate more than 50 percent of the earth's land surface already has been altered by people. Worse yet, they say, human activity is accelerating the rate of transformation at such a speed that it's not sustainable.

This disruption happens every time a wetland is filled or drained, a ridge leveled, a forest chopped down, natural vegetation ripped out, and lowlands filled or flooded. Too many disruptions from land use change can harm nature's ability to recharge a water area and can lower water tables. At times, they can also cause flooding, if developers have ruined natural drainage patterns.

As the world becomes increasingly smothered with impervious asphalt and concrete surfaces, a greater amount of rainfall becomes stormwater runoff. When this happens, the volume of groundwater recharge and evapotranspiration decreases.

"With intense urban development," says Tom Singleton, a Tallahassee-based water consultant, "there can be a fivefold increase or more in the amount of runoff, an 80 percent loss of deep infiltration, a 60 percent loss of shallow infiltration, and a 25 percent loss in evapotranspiration."

Some water loss can be tolerated. But too much causes distress to all living things. When water experts talk about water sustainability, they mean there's a point at which no more water should be taken from groundwater, because it is not being replenished, or recharged, fast enough by rainfall to keep up with human pumping.

But there's another big problem. Land use changes also change the profile of water, which is all too often a destructive act.

What makes a water area its own distinct self is its "hydrograph," explains Mark Rains, an ecohydrologist at the University of South Florida. Imagine, he suggests, graph paper with an x and y axis. Water may stand in a certain place for a period of time, as represented by coordinates on the graph, and at another place when it dries, shown by two other coordinates.

This distinct "wiggle" shows its profile and gives the water area an identity. "Changing an area's hydrologic system has a consequence," Rains warns. "That cypress dome may look the same to the casual observer after the water's been taken away—or even if there's an influx of more water, or a change in water quality—but things are going on between water and the biota."

Another issue deals with what you might call "hydrologic damage yet to come." Designing water infrastructure to minimize environmental damage when you're building a new development is hard enough, but it's trickier still to get it right for the long run. That's because conditions are going to change, especially in hot, humid, hyperactive Florida. What's here now will be gone—or greatly changed—in fifty years. That's true for both the developments and ambient environmental conditions. Planning for those coming changes should be in bold print in every Florida water supply plan.

Some water experts think Florida's hydrology—a term often used to describe the natural movement, distribution, and quality of water—has taken too many hits. "We've passed a tipping point," laments Tom Singleton, "and it's not just because we've exceeded our capacity. It's also because we've changed the hydrology. We've reduced the native capacity of the land form to capture, store, and cleanse water. Figuratively speaking, we're standing on our own water hose."

The future doesn't bode well. "If Florida does not significantly alter its policies regarding growth and development, more than 7 million acres of additional lands will be converted to urban uses by 2060," according to the 2017 Water Report included in "Water 2070" from 1000 Friends of Florida, a not-for-profit smart growth advocacy organization, along with the Florida Department of Agriculture and Consumer Services (DACS), and the University of Florida Geoplan Center. "This means that Florida will lose 2.7 million acres of existing agricultural lands, along with 2.7 million acres of native habitat. And more than 2 million acres within one mile of existing conservation lands will be converted to an urban use."

The report also warns, "If we don't change the way we are developing, more than 1/3 of Florida will be paved over."

Millions More on the Way

More land use changes are also coming to Florida if all those newcomers actually do arrive, and that's bad news for water supplies. To grasp the scale of Florida's demographic change that's already occurred, consider that the state's population increased from about 2.7 million in the early 1950s to more than 20 million in 2018—a tenfold increase, making Florida the nation's third most populous state.

In 2017 the pro-business Florida Chamber Foundation, the research and solutions-development arm of the Florida Chamber of Commerce, came up with its own prediction, which anticipates the state growing by an additional 6 million permanent residents over the next twenty years. "That is the equivalent of the entire population of Maryland packing its bags and moving to the Sunshine State," observes David Childs, a lobbyist for the Florida Chamber of Commerce.

The foundation may be lowballing the number of expected transplants. The 1000 Friends of Florida "Water 2070" report predicts the arrival of 15 million new permanent residents in the next half century.

And there are still others on the way to Florida not included in those estimates—tourists who already outnumber residents five to one; their numbers are expected to grow, too.

Some Floridians seem pleased by this news. "I'm thrilled that Florida continues to break records in the amount of tourists visiting our state," Florida governor Rick Scott announced in May 2016. "Last year, we exceeded our goal of an historic 105 million visitors because of Florida's unique national treasures, and I look forward to smashing our record again this year by welcoming 115 million visitors to the Sunshine State."

Scott, however, had nothing to say about where the extra water will come from to accommodate the newcomers when they arrive.

And there just may not be enough. According to the water report, if Florida plunges ahead with a business-as-usual approach to development, water demand in cities and suburbs will increase by 106 percent over the next half century.

Using 2010 data as a baseline, the report also shows Floridians use about 5.27 billion gallons of water every day. This figure could rise to 8.1 billion gallons per day by 2070, when the state's population may have increased by 75 percent.

But there's something else in the report worthy of a second look: the presentation of an alternative track the state could take instead of the one it seems to be on now. Through a mix of conservation methods and more compact construction practices, demand for water in Florida in 2070 could be reduced to 6.85 billion gallons per day, a rise of 30 percent over the baseline level of 2010.

That means, however, there'd still be a demand for at least a third more water than there is now.

Sadly, even that figure may be an understatement. "The study doesn't mention how much water is being wasted going to the ocean," says Gary Goforth, a thirty-year veteran environmental engineer in Stuart,

Florida. "Nor does it identify the water that's needed for restoration projects, like the water that's needed to be sent south to the Everglades."

It also excludes other troubling factors such as water contamination, evaporation, and damage done to the state's hydrology from intense urban development.

Nor does the study consider sea level rise, which climatologists expect to play havoc with Florida's weather, hydrology, and aquifer—all of which affect fresh water supplies.

Rain replenishes most of Florida's aquifers. But no water planner is suggesting that rainfall is expected to increase to keep up with the water needs of the millions of newly arriving people into Florida.

Small wonder many water management officials are looking for ways to find "new water" sources.

A Limited Supply

Is there any new water to be found? Not really. Scientists have long known that 97 percent of the water on earth is salty. Another 2 percent is fresh but unusable because it is locked up as glaciers or other forms of ice, or it is embedded in the ground—much of it too hard to get. The remaining 1 percent is also fresh, but only 0.3 percent of that amount is suitable for humans, and most of that quantity is tucked away underground in aquifers that often are not recharged quickly. Some fully confined aquifers aren't recharged at all. Once tapped out, they are gone from that location forever. The rest of the water ends up in the atmosphere, living things, and the soil.

That small amount of fresh water available to humans is therefore essentially finite. The good news is that it also can't be used up. It is the same water that's been here for millions of years. Dinosaurs slurped some of the water molecules that Americans drink today. Thomas Jefferson drank traces of the water that finds its way into the plastic water bottles hoisted by Virginians today.

Thus, roughly the same volume of water exists today as in the distant past. In fact, there could be, perhaps, a little more. Some scientists think new volumes of water are introduced to the earth every day as huge, hurtling ice balls the size of small houses from outer space turned to water vapor after striking the earth's atmosphere. Perhaps as much as three trillion tons of space water are added to the world's water supply every ten thousand years. The trouble is there aren't enough of these alien snowballs to increase it "fast enough to solve humanity's water shortage problems," explains E. C. Pielou, a Canadian statistical ecologist.

When water policymakers talk about developing new sources of water, they often mean they are seeking new access to existing water supplies. Dipping into the St. Johns River for potable water could be one of them. Tapping underground seepage could be another. Other "alternative water sources" include conserving water and purifying brackish groundwater, seawater, and stormwater.

Increasingly, this goal also means finding new ways to reuse water, as many new advanced treatment technology companies are doing. In one sense, recycling water is nothing new. It's always being reprocessed through the water cycle: groundwater, surface water, evaporation, rain.

Water is always transforming and turning into one of three forms: gas, liquid, and solid ice. It flows in many directions and speeds. Sometimes it rushes and on other occasions it creeps across the surface of the earth or seeps beneath it. Water flows underground through layers of rock. Obeying the law of gravity, much of it drains into the sea. It moves vaporously skyward, too, either through evaporation or transpiration from the tiny pores of leaves. In short, all this complex movement, shape-shifting, and distribution of water is the hydrologic cycle.

Nobody has inexpensively made from scratch quantities of brand new fresh water on a scale large enough to be used as a reliable water supply. The closest thing is the much-touted practice of desalination—that is, the process of getting salt and other impurities out of sea or brackish water. But it is costly and energy intensive.

Unbalancing the Water Budget

Finding enough water to satisfy consumer demand isn't the whole problem. There's also the "water budget" to contend with. Water experts often use an economic metaphor to explain the idea. If you want to determine, for example, how much you can spend on Christmas gifts, you need to know what your budget is. This means taking inputs and outputs into consideration. First, you'd write down your input—your income. Next, subtract fixed expenses, or outputs. This covers fixed costs like insurance and rent. The amount that's left over can be stored in a bank or spent.

A water budget works in a similar way. Rain that infiltrates the ground is an input to the aquifer. Outputs include human activities such as extracting water from the aquifer for lawns, agriculture, and flushing toilets. There are natural outputs too, like evaporation or a discharge into a river. Water that is left over is stored in the plants and the ground. If there's too much water, flooding and drainage problems will occur. Too little water means plants will dry up. When the input and outputs are about the same, the hydrologic system is in equilibrium.

Trouble arrives, though, when water is extracted faster than the recharge to a water area, which includes surface water and groundwater—often called *blue* water—plus the water stored in all the surrounding vegetation—or *green* water. Maintaining the volumes of both green and blue water is essential to making sure there's enough water to meet the needs of humans, other life forms, and natural systems.

"The quantity of water in each segment of the cycle—the water balance—defines the local landscape: the type and amount of vegetation; the type and prevalence of streams and surface water features; and the amount of water in the groundwater aquifer," explains Singleton. "The natural landscape has evolved with the water balance of the watershed—with vegetation, soil, and aquifer absorbing water where it falls and flows."

If Singleton is right, too many water planners may be missing something. Trying to solve the state's water quantity problems by seeking new water supplies is only a partial solution.

What's more important—and harder to do—is to save and restore the state's natural hydrology.

All these issues—as dire as they may seem—don't mean that Florida's water supply problems are intractable.

In fact, "There are lots of solutions to water quantity," says Janet Llewellyn, a former policy administrator for the Office of Water Policy at Florida's Department of Environmental Protection. "We're not out of water; we're out of cheap water."

Then she adds, sounding a little less confident, "But water quality? Now, that's a much harder problem."

2

Fouling the Waters?

FLORIDA'S DRINKING WATER IS "generally good," according to a report from the Southern Regional Water Program, a partnership between the U.S. Department of Agriculture and land grant colleges.

That sounds reassuring, until you read the next sentence: "But local groundwater contamination problems exist." And that's something to think about, considering that more than 90 percent of Floridians get their drinking water from groundwater sources.

Besides, not everyone agrees that the water quality of Florida is all right. Public Employees for Environmental Responsibility (PEER)—a national nonprofit alliance of local, state, and federal scientists, law enforcement officers, land managers, and other professionals that functions as a watchdog for possible violations of environmental law—for instance, claims there are drinking water quality problems galore in Florida. The title of its August 2016 report tells just how serious PEER thinks the situation is: "Don't Drink the Water: Collapse of Florida's Safe Drinking Water Enforcement Program."

Among PEER's more disturbing findings is that nearly 700 of Florida's estimated 5,300 public water systems don't comply with safe drinking water rules. Enforcement is lagging, too, according to the watchdog group. "Despite nearly 2,000 individual safe drinking water violations, including 295 violations involving exceedances of maximum contamination limits for things such as total coliform [a group of bacteria used as indicators of pathogens in water treatment supplies], chemicals, radionuclides and disinfection byproducts, the DEP opened only five enforcement cases in 2015 and assessed relatively small fines in only two."

Moreover, PEER claims the number of penalties has plummeted since 2010, "with the number of assessments in potable water cases dropping from 141 to only 2 (a 98% nosedive) and the total amount of fines assessed plunging from roughly a quarter-million dollars to a mere $12,000 (a more than 95% falloff) in 2015."

These findings reveal "Florida no longer has a functioning program for ensuring safe drinking water," asserts Florida PEER director Jerry Phillips, a former DEP enforcement attorney who today analyzes raw data obtained from his old employer.

However, Janet Llewellyn, who worked at DEP when the PEER report was issued, cautions the situation is not as alarming as Phillips claims. Some of the cited infractions were minor, she says, and many missing reports of violations were merely tardy, not absent. "And Jerry," she says, "might have been putting more emphasis on some infractions than they deserved."

But she does concede, "There have been budget cuts in recent years which have made it harder for DEP to do its job."

Phillips, however, thinks the problems are older than current budget cuts. He worked at the department during the mid-1990s for the Lawton Chiles administration, when the state merged the Department of Environmental Resources and the Department of Natural Resources, creating the Department of Environmental Protection. "I was part of the negotiations to get delegated authority from the EPA to handle the national pollution discharge program for Florida, which dealt with the permitting and enforcing of pollutants being discharged into surface waters," says Phillips. "EPA Region 4 in Atlanta didn't have the manpower to properly administer the program, but it could turn the authority over to Florida, which could control events with administrative oversight."

First, Florida lawmakers had to create legislation to allow the transfer. However, says Phillips, "Big-time polluters were almost writing this legislation, and they also had a hand in crafting the actual water pollution rules and regulations. It was a blank check for them."

After a while, Phillips's morale sank, and he suspected many of his colleagues felt the same way. "During those years," he says, "there was a continuous exodus of scientists from the DEP who said their work no longer had anything to do with environmental protection, but instead had everything to do with feathering the nests of the big-time polluters. They were replaced with people with no experience, who did what they were told and believed they were doing the right thing. The level of fear at the agency was so bad that employees would drive away from where they worked and only talk to me on their cell phone because they were convinced that there was eavesdropping going on."

Finally, it was all too much for him. "I left the DEP, because the writing was on the wall. I was once handling fifty new cases a year; now I was down to two. It became obvious that I was just taking up space. It was a very depressing time."

Phillips next worked at the general counsel's office at the Florida Department of Transportation. In 2003 he went to work for PEER, "with all this knowledge about the DEP." Right away he started submitting records requests for enforcement cases as far back as 1987. Using that information, he has been publishing annual reports on the DEP ever since. "They know I know, and they don't like it."

More Bad News

PEER, however, isn't the only group to report problems with Florida's drinking water. In 2009 the *New York Times* invoked the Freedom of Information Act to obtain water pollution records from all fifty states and the federal Environmental Protection Agency (EPA). Next, it put together a national database of water pollution violations, which, the *Times*'s editors claim, "is more comprehensive than those maintained by states or the EPA."

Their research showed the federal Clean Water Act had been violated in excess of 506,000 times between 2004 and 2009. According to researchers 35,300—or about 7 percent—of those events took place in Florida. These figures, says the newspaper, are based on reports filed by the polluters themselves. Nobody knows how many violations were not reported.

In response to the newspaper's request for comment on its finding, the DEP complained the EPA data doesn't tell the whole story. For instance, says the agency, it doesn't reveal how many "chances" violators are given. Nor does it make clear that some violations are resolved "without the need for formal enforcement."

More recently, Florida appeared in the national spotlight for having lead pollution problems. This revelation came in the wake of the 2014 drinking water crisis in Flint, Michigan. City water managers of the northern industrial city had decided they could save money by quietly switching to a new water source, the Flint River. But the new water had old pollutants that reacted with water pipes, culminating in lead contamination that exposed children to a heightened risk of brain damage and adults to maladies such as hypertension and kidney problems.

Since Flint's water catastrophe, other U.S. cities and communities have discovered to their dismay that they have lead problems, too. In a March 2016 analysis of the federal Environmental Protection Agency, the *Florida Today* newspaper concluded nearly two thousand water systems in the United States had elevated lead levels in their drinking water.

Florida made the list. According to the analysis, almost fifty thousand Floridians may have been exposed to dangerous levels of lead in their drinking water between January 2012 and June 2015. Moreover, 10 percent of the test samples taken from sixty-four Florida water systems during that period showed a lead level exceeding the federal standard of 15 parts per billion. At this level, water utilities are required by the EPA to alert the public of the danger to their health and take measures to fix the problems.

In May 2017 came more alarming news. The Natural Resources Defense Council, a New York City–based nonprofit international environmental advocacy group, released "Threats on Tap: Widespread Violations Highlight Need for Investment in Water Infrastructure and Protections," a study revealing all fifty states, plus the District of Columbia, Puerto Rico, and other U.S. territories, had been adversely affected by violations of the Safe Drinking Water Act, including problems with disinfectants, lead, copper, and coliform.

Based on comparisons of the size of populations served by water systems that were in violation of the act, Florida ranked second highest with 7,540,465 people.

Meanwhile, unease was emerging across the state over a new DEP proposal to relax restrictions set by the EPA on scores of pollutants discharged into Florida waters. The state agency insisted the proposed rule changes were scientifically derived and would still provide safe water. It also claimed that thirty-nine extra chemicals would be monitored.

In the wake of widespread criticism, DEP secretary Jon Steverson told the *Tallahassee Democrat* that news media had "inaccurately and unfairly" depicted the agency's proposal. He also asserted Florida "has some of the most comprehensive water quality standards in the country, including the most advanced numeric nutrient criteria in the entire nation."

Critics of the plan, however, couldn't understand why the state's main water protection agency was backing off—instead of toughening—pollution rules, especially as Florida's population was expected to skyrocket.

One of these critics is Randie Denker, a Tallahassee-based environmental attorney and cofounder of Water without Borders—a nonprofit organization seeking to supply fresh water to people around the world. She was also one of DEP's first enforcement attorneys. "Prior to late 1980 or early 1981," she recalls, "each attorney at DEP was assigned a region. I handled the entire southeastern region, which included Dade, Broward, Palm Beach, Martin, Okeechobee, and a few other smaller counties. In their region, each attorney handled all sorts of cases, both permitting and enforcement." Not long afterward, "DEP set up a unit solely for enforcement and nothing else. I made it known that I would like that job, and I was the first attorney hired to fill the position."

Today she is hard on her old employer. "DEP's new proposed criteria will lower standards for 89 known carcinogen and/or human health–based toxic chemicals," Denker warned in a January 9, 2017, personal letter to all state senators. "Floridians will get weaker protection from almost two dozen carcinogens. Essentially all of the chemicals would be allowed in our drinking water supplies, shell fishing areas, swimming and fishing waters at significantly higher amounts."

Other skeptics charged the rule change was meant to pave the way for the legitimization of chemicals, such as benzene, used in fracking—the controversial process of injecting water at high pressure deep into the earth to extract oil or natural gas. So far, the effort to legitimize fracking hasn't made much headway in the Sunshine State.

Controversy only intensified when Florida's Environmental Regulation Commission (ERC) met in Tallahassee in July 2016 and approved the DEP's rule change by a 3 to 2 vote, prompting a howl of public protest. Many critics griped that the council was supposed to have seven, not five, members. One of two empty seats was required to be filled by someone from the environmental community, and the other by a person representing local government. Governor Rick Scott, despite urgings from around the state, kept both seats empty. By doing so, he helped to determine the outcome of the vote.

Denker remains annoyed not only by the ERC decision, but also by the way Florida goes about creating its water pollution rules. "The ERC is composed of people who supposedly represent seven different sectors, including the public," she said. "They are the ones who put the imprimatur of the state on the final decision. But no one on the council has to have any scientific credentials. There's no oversight. Nobody elected them. And the whole process is highly political."

Another Floridian who objects to the way the ERC does things is John Moran, who at the time of the rule change controversy was a twenty-six-year-old Tallahassee native, working on a PhD in anthropology at Stanford. He attended the July ERC meeting. A moment after the decisive vote was taken, Moran dashed to the dais, occupied one of the empty seats and exclaimed, "The environmental community hasn't been given a voice. The local governments haven't been given a vote."

His protest sparked cheers in the audience before he was escorted from the hearing room. "I was appalled by the weakening of our state's restrictions on more than two dozen carcinogens in our water—including benzene, a chemical used in fracking," Moran recalled later. "Why are we using a model no other state uses to loosen standards on carcinogens as the state faces its worst water crisis in recent history?"

Orlando-based State Senator Linda Stewart (D) is also an ERC critic. "Three people making a decision on the pollution levels of drinking water for the whole state—how absurd is that?" she asks. By the end of 2017, Stewart was trying to advance a bill through the senate requiring the governor to appoint all positions to the ERC within ninety days.

The Seminole Tribe took action too. But its lawsuit challenging the new water rules was dismissed in court when the tribe missed the deadline for petitioning for an administrative hearing on the new rules—by two minutes.

Water Pollution: An American Tradition

Contending with unsafe drinking water isn't new, of course. More than two centuries ago, in fact, many American colonists put up with notoriously unclean water supplies. Water drunk from many swamps or stagnant lakes made people sick, or sometimes killed them. Diphtheria and typhoid were ever-present in dirty water.

Spring water was often considered the purest, but it was generally located in the lowlands, where diseases like malaria and swarms of insects lurked. That's why many settlers preferred instead to build their cabins on safer, higher ground, even if it meant they had to haul their water uphill.

Many colonists also had to contend with impurities in shallow brackish wells and streams contaminated by human waste and garbage. Nonetheless, it was possible to raise taxes to clean up these suspect water sources, writes Eric Burns, author of *The Spirits of America: A Social History of Alcohol.* However, he suggests the nation's early settlers tended to avoid such actions: "Why pay extra money to sanitize a beverage that no one wanted to drink in the first place? Why not just give up on the stuff and spend smaller amounts of money on safer, more stimulating drinks?" Thus, many Americans instead relied on drinks such as hard cider, rum, and beer—"ardent spirits" that they believed warmed them and contained medicinal properties. Water was for animals. Alcoholic drinks were gifts from God.

Attitudes changed, however, as cities grew and public water systems became more important. All too often water quality grew worse as the Industrial Revolution slammed into high gear during the second half of the nineteenth century, turning once pristine American rivers black, nasty, and polluted with raw sewage and industrial waste. Cities were installing sewers, but these systems frequently dumped human filth directly into rivers and estuaries where it often made its way into water supplies. It wasn't uncommon for nineteenth-century Chicagoans to turn on their water spigots at home and see dead fish and human excrement come out. Due to poor sanitary conditions in city reservoirs, water-borne diseases such as cholera and typhoid fever took the lives of many Americans.

Personal habits contributed to the problem, too. Sharing a single, community drinking glass, for example, was the norm at many drinking fountains throughout the country, as it still is in many places of the world. Part of the problem was simple ignorance. It wasn't until around 1900, in fact, that most American doctors got around to accepting the "germ theory" that microbes cause disease—which even school kids take for granted today—and began washing their hands before surgery and sterilizing their instruments.

Attack on Dirty Drinking Water

About this time another notion—that public drinking water should be purified—caught on, too. In 1908 physician John Leal worked as an advisor to a water company charged with cleaning up the contaminated water supply to Jersey City. Leal, whose father had died from contaminated water in the American Civil War, believed the company's solution of building new sewers was useless. So, relying on the pioneering work of others in Europe and Chicago, Leal instead secretly enlisted the help of an engineer to build a system that added dissolved chloride of lime to New Jersey's Boonton public water supply reservoir for two hundred thousand people.

It was a dangerous gamble, because at the time there was great public hostility to adding chemicals to public water sources. But Leal's plan worked stunningly. Illness and death from typhoid fever for those using the city water nearly vanished. Soon, other cities were following Jersey City's example, and the number of deaths from typhoid and cholera decreased dramatically across the nation. Harvard professors David Cutler and Grant Miller have determined that by 1930 chlorination and other water infiltration systems had cut infant deaths in American cities by 74 percent and for children by about the same amount.

The people of Jersey City haven't forgotten Leal's accomplishment. On September 24, 2008, they celebrated "100 Years of Chlorination."

Even more progress was made in the 1970s, especially with the creation of the EPA, along with a slew of other public safety state laws, local regulations, and federal legislation such as the Clean Water Act and Safe Drinking Water Act. All these helped make drinking water in Florida and many other U.S. locales—despite their current problems—some of the best in human history.

Many people around the world aren't so lucky. "Daily about one out of six people on earth lack dependable access to clean water. Nearly

1,000 children around the world die each day from waterborne illnesses," according to UNICEF, the United Nations program for needy children and women in developing nations.

Unseen Perils

Florida lacks the bleak, industrial landscapes that mar skylines elsewhere in the world, but that doesn't mean it doesn't have water quality problems. For instance, ninety-two federal superfund cleanup sites pollute the state, as do raw sewage spills in Naples, St. Petersburg, and many other municipalities. Radioactive water from power plants in Miami and phosphate mining residue in Polk County possibly are wending their way toward underground fresh water supplies. Arsenic sludge was recently detected seeping into groundwater in Ft. Myers.

Infrastructure Blues

Contaminants, however, aren't the only challenges water managers face as they try to keep Florida's water safe to drink. One of the biggest headaches is overlooked because it's hidden from view—the state's water infrastructure. In 2015 the EPA announced that Florida will need to spend $16.5 billion over the next two decades on pipes and other hardware merely to keep providing safe and reliable water to its residents. Such an expenditure would be part of a $1 trillion undertaking to make freshwater infrastructure overhauls nationwide.

There's also a growing concern over the vast underground systems that carry away human waste. "Florida's wastewater system is increasing in age, and the condition of installed treatment and conveyance systems is declining," according to the 2016 Report Card for Florida's Infrastructure, sponsored by the Florida Section of the American Society of Civil Engineers. It's high time, say the engineers, to invest in Florida's wastewater infrastructure.

Ft. Lauderdale is one city that can no longer ignore its water disposal problems. By 2017 it was under state scrutiny, which meant it had to pay fines and $117.5 million in mandatory system repairs and improvements after its aging pipes had spilled nearly 21 million gallons of raw sewage into local waterways and groundwater in the previous three years.

The state also will need $1.1 billion in capital improvement just to keep its stormwater operations going until 2019, according to the report card, which also points out that "only 1 in 4 stormwater utilities stated that today's operation and maintenance capabilities were adequate only to meet the most urgent needs."

Skeptics, of course, wonder if these claims are exaggerated by engineers who may have a self-interest in new infrastructure construction. Nonetheless, it's only reasonable to assume the state's current massive systems of rusting, leaking pipes, valves, pumps, locks, and so forth will be under extra stress if fifteen million new residents want to use them.

"We have communities in Florida with 100-year-old pipes still," Tom Friedrich, a Jacksonville-based water infrastructure expert, told the Associated Press. "Old pipes cost more to repair now, and it's got to be done over many years. It's a real headache."

Florida's flat terrain has always made disposing of unwanted water a problem, but having aging pipes and equipment only makes things worse. The latest reminder of this fact arrived in September 2017, in the wake of Hurricane Irma, whose furious behavior caused problems for wastewater and stormwater systems across the state. Hundreds of millions of gallons of dirty water either leaked from bad pipes or backed up and spilled out of manholes or other openings and then flowed into freshwater bodies and across land surfaces. Sometimes these mishaps were the results of power outages at wastewater facilities that prevented pumps from working. One-third of Tampa's 230 sewage pump stations, for example, lost power during the storm.

Neighboring St. Petersburg struggled with four major spills during Irma's visit, the biggest of which sent 430,000 gallons of partially treated wastewater out a treatment plant holding tank.

Raw and treated sewage also spilled in the streets of the Central Florida city of Oviedo, while Jacksonville registered more than 2.2 million gallons of wastewater spillage. Six million gallons of partially treated wastewater poured into Miami's Biscayne Bay.

Once the storm had moved on, officials at more than a hundred water systems in forty Florida counties advised local populations to boil their tap water before consuming it.

Such failures threatened fresh water supplies and had to be reported to the DEP, which has the responsibility of upholding and enforcing federal water quality standards. It's a big job. In fact, the department has to keep track of 5,275 active public and private drinking water treatment systems in Florida.

The DEP is also responsible for the quality of Florida's water bodies. Grasping the extent of any water pollution, however, is tough, because there are so many of these water bodies in Florida. Not all are regulated or monitored. But a sample of the overall health of Florida's waterways that have been checked out appeared in a November 13, 2012, study, "Valuing Florida's Clean Waters," prepared by the Stockholm Environment Institute, an independent, international research organization and research affiliate of Tufts University in Massachusetts. Among other things, the institute found that 5,587 river miles, 919,000 lake and reservoir acres, and 1,795 square miles of estuaries had "nutrient-related impairment—meaning that they are no longer clean enough for a specified use."

Sources of Pollution

Not all water pollution is caused by humans. Nature also produces plenty of substances toxic to humans. Arsenic, for instance, a deadly residue of industrial activity that often leaches into Florida groundwater, also occurs naturally at several locations on the west side of the state, ranging from Dixie County to Hillsborough County and the western half of Polk County. Too much of it in a water supply can be dangerous, which is why the DEP recommends testing wells in areas where arsenic is known to be a problem.

Natural erosion is another cause of water pollution. The EPA estimates that it is responsible for nearly 30 percent of all sediment in the United States, though the other 70 percent—so-called accelerated sediment—comes from erosion caused by human land use. Soil that's swept away by flood waters often releases naturally occurring nitrogen and phosphorous, which feed algae in the water. All sediment pollution causes $16 billion annually in environmental damage nationwide.

Shifting conditions in the natural world also affect water quality. For instance, a moving river has more dissolved oxygen—which aquatic creatures need—than a stagnant lake does. The time of the year matters, too. In winter, dissolved oxygen levels tend to be higher than in summer, when bacteria are more likely to thrive and busily decay organic matter.

Violent storms can also mess up water quality. Hurricanes, tropical storms, and other intense rain events often blow in clouds of polluted dust and water particles, along with exotic seeds, microbes, insects, and even birds. All of them eventually interact with Florida waters.

Natural pollution is hard enough to keep up with, but it's the anthropogenic, or human-caused, variety that often commands attention the most. Such was the case in the early 1980s, when four hundred drinking-water wells in northeastern Jackson County, Florida, were identified as contaminated by the pesticide ethylene dibromide.

Of the seven hundred or so new chemicals introduced into the natural world every year, it's not always clear right away which ones are harmful and need monitoring. And even when it is, regulators are often clobbered by budget cuts or intimidation and pressure from elected officials not to interfere with polluters.

Figuring Out Whom to Blame

Before Florida's water regulators can monitor and maybe even control pollution, they try to figure out where it came from. If they can pinpoint, say, a certain factory or wastewater treatment plant, they've detected a "point source."

Over the years, Florida's water quality enforcement officials have gotten better at tracking down point source polluters. A big reason for this is today many polluters must get permits for their discharges, which makes it easier to keep track of what they're doing.

Non–point source pollution is trickier. This is pollution caused by many people. Stormwater, for example, is loaded with sediments that come from multiple sources. Imagine fallen rain turning to stormwater as it races over streets, highways, parking lots, factories, landfills, golf courses, illegal dumps, construction sites, and backyard lawns. Along the way, it picks up oil, grime, soil, leaves, metal fragments, litter, wood fragments, animal waste, poisons, pesticides, insecticides, birth-control pills, antibiotics, suntan lotions, and agricultural chemicals.

If untreated, this filthy cocktail will discharge into a water body, such as a lake, bay, or river, where it can wreak widespread, serious ecological damage. It can alter water chemistry and smother and poison flora and fauna. Sometimes the contaminants increase turbidity, which makes it hard for fish and other aquatic creatures to see. Sediment often ends up coating the gills of fish or, if swallowed, damages their organs and tissues. Reproduction of aquatic creatures is sometimes affected, too, when pollutants affect their genes. In 1998, for instance, University of Florida research teams discovered that pesticides and other chemicals were causing reproductive and hormonal problems in alligators in Lake Apopka and Lake Okeechobee, such as reduced testosterone levels and smaller penises in male alligators. "This should be a wake-up call," UF professor of zoology Lou Guillette warned at the time. "We have to make sure that similar problems are not occurring in ourselves."

But he later concluded humans were indeed victims, too. In the 1990s, he appeared before Congress, warning that "every man in this room is half the man his grandfather was."

Florida's flat terrain and its hydro-geologic features make it susceptible to groundwater contamination. In many places, rain quickly soaks through thin or sandy soil, taking along with it any contaminants as it percolates porous limestone lying underneath. Unless it hits a layer of clay, it can swiftly reach the water table, which tends to be relatively high in much of Florida.

Polluted water can also enter the aquifer via human-made storage systems. As a report from the Southern Regional Water Program—a partnership of USDA NIFA and Land Grant Colleges and Universities—points out, "There are tens of thousands of potential point sources such as surface-water impoundments, drainage wells, underground storage tanks, flowing saline water wells, hazardous wastes sites, power plants, landfills and cattle and dairy feedlots."

Floridians' penchant for living near water makes the problem worse. In the Panhandle, says Laurie Murphy, executive director of the Emerald Coastkeeper in Pensacola, many residents build their houses on the banks of the region's plethora of scenic creeks, rivers, and streams. The concrete and asphalt driveways at their homes, however, "destroy habitat for animals," she points out. "In addition, the natural grasses, trees, and shrubs are no longer there to absorb the nutrients from fertilizers and septic tanks."

All this facilitates soil erosion, and the more of it there is, the faster the stormwater flows into the water, she adds. "It creates a wider girth all the time that affects everything."

This phenomenon is not unique to the Panhandle. It happens across the state wherever Floridians build along the banks of canals, rivers, wetlands, and lakes. Around the clock, seepage from leaky and poorly maintained septic tanks, along with storm-runoff loaded with pesticides, fertilizers, and fungicides from lawns, flush into nearby water bodies or seep into an aquifer every time it rains or sprinkler systems click on.

Murphy, armed with such knowledge, along with a degree in oceanography and a master's in geographic information science, wants to awaken others. Not long ago, she launched a series of podcasts about water quality and riparian issues. "I started it as a way for the community to get facts from scientists with real-life experience. So far I've received a lot of positive feedback."

Still, it's hard to stay upbeat. "There are lots of problems," she says with a sigh. "We have superfund sites, a coal-fired plant that's leaking arsenic in the Escambia River, along with stormwater problems. I was out in Carpenter Creek not long ago and just wanted to cry, looking at the garbage piles stacked up high; it looked like a third-world country. We should be blaming ourselves, not water utilities."

There is some good news, though. "All new construction has to have stormwater treatment; it's been that way for decades," says DEP's Janet Llewellyn. "The problem is that a lot of Florida was built before the new requirements were put in place. Still, a lot of stormwater is being treated, and anything is better than what we once had."

Nonetheless, detecting and removing pollutants is more expensive than preventing the foul stuff from getting into water in the first place. "Our environmental structure is reactive, when we're dealing with environmental pollution," suggests Heather Obara, an environmental attorney and associate director of the Howard T. Odum Florida Springs Institute. "We're always thinking about it after it's occurred. Rather than being proactive and taking steps to protect our environment, we're dealing with it on the back end."

The situation is even worse if Florida's environmental laws are not being vigorously enforced, she says. In that case, the laws are useless.

Meanwhile, more than twenty million Floridians—and hundreds of millions of tourists—are relentlessly heaping into fresh water bodies unknown quantities of thousands of contaminants. Among the substances are those "feeding" some of the most worrisome of all pollutants. They are alive, growing, and reproducing.

3

What's Nasty, Deadly, and Green All Over?

THERE'S PRIDE IN Fred Rossiter's friendly voice when he says he's a certified charter boat captain. Not that he discounts his other former selves: realtor, property appraiser, and contractor. But what this Ohio-born, longtime Colorado resident has always loved to do above all else is to fish. And what better way to make a living, he often asked himself, was there than to take people fishing off the Florida coast?

In 2003 Rossiter tried to fulfill his dream by moving with his wife to Pine Island, off the state's southwest coast, to live in a marine environment. He loved everything about his new home—island life, boating, fishing, birds, and especially "all the critters in the water." It didn't take long for him to obtain his captain's license, and soon he was taking fishing tourists out to the gulf aboard his twenty-foot boat, *Sweet Sue*.

At last, life was good for Fred Rossiter. He was a successful charter captain and having fun. "I was taking out fifty charter trips a year, making $60,000 a year," he recalls. "That was good enough for me."

Then came the troubles. The first was the 2008 Great Recession that wrecked the state and local economies. In 2010 the BP Deep Horizon oil spill followed. "The oil never got down here to the southwest coast," says Rossiter, "but the public perception outside the state was that it had, and lots of tourists stopped coming."

Next to arrive were two winter cold snaps that killed so many fish that Pine Island marina officials had to haul them away to clear out the waterway to the island.

Then came more fish kills, Rossiter says. Some occurred on the coast and were caused by the Florida red tide—a naturally occurring harmful algal bloom that's appeared now and then for centuries. Fish also died from a different kind of algal bloom that flourished on the polluted Caloosahatchee River, which drains the northern edge of the Everglades and enters the Gulf of Mexico ten miles southwest of Ft. Myers.

When Rossiter first arrived in the Charlotte Harbor area, he could see flocks of birds spiraling over the gulf. "The water would be boiling with bait fish. You'd see thousands of fish, thousands of rays, along with hundreds of small wading birds."

But now they were almost gone—or at least not as plentiful. When Rossiter voiced his concerns to state wildlife officials, he says, he was told there was no data to back up his claims. He was infuriated. How could anyone say that when all you had to do was look at the water and see for yourself?

Rossiter wasn't the only one to witness the decline. Others saw it happening too, he says, including members of the local tourist industry and some fishing guides who told him to keep mum, so he wouldn't scare off tourists.

They found out anyway. "I used to take three or four people out on a charter cruise, and we'd catch ninety fish a day—all of them keepers for dinner," says Rossiter. "But three years later, when I took fishermen out, we only pulled in about two dozen, and a lot of them were trash fish."

Making matters worse, he was now down to making just one charter trip a year. But in many ways that was one too many. Rossiter says his conscience wouldn't let him take any more people out. When prospects from up North called him to book a trip, he advised them not to come to Florida. "It wasn't worth it," Rossiter told them, "because the fish are gone."

At that point, he realized, "Staying here doesn't make any sense. I'm seventy-four, and all these things have taken the passion out of my reason to be here. The big draw for living here in Southwest Florida no longer exists. For a while it was a lot of fun; that was before everything went to hell. By the time anyone fixes this problem, I'll be dead."

Rossiter sold his boat. In March 2017 his house was on the market.

It was time to spend more time with grandkids back in Colorado. But even with that blessing, he says, he would have stayed in Florida if things had been different.

Rossiter is not the only one to quit the charter fishing trade. A half dozen or so of his local captain friends had already left the state. Rossiter says with a sigh, "I'm the last one."

But he's not really the last man standing. Others are still fighting. Among them are Captains of Clean Water, a group of angry fishing guides from Ft. Myers who say they've "had enough" of the state's water management practices. They want an end to the destructive, algae-ridden Lake Okeechobee discharges into the Caloosahatchee and St. Lucie River estuaries.

In April 2017 another group of salts—forty Florida Keys fishing guides and their allies—drew the attention of the news media when they arranged their skiffs in the shallow flats near Islamorada to spell out a one-word plea readable from the sky: "Help." It was the anglers' way of focusing attention on the reality that the lack of fresh water was hurting their fishing grounds and their former multimillion-dollar industry.

A month earlier, business leaders from a major fly-fishing industry, a boating company, and tourism officials convened with Governor Rick Scott and lawmakers to complain that a major cleanup of polluted outflows from Lake Okeechobee was needed for environmental protection and economic reasons.

What upsets these Floridians causes distress statewide. It reminds people of an old monster movie.

Modern Blobs

Farfetched? Maybe not. Think back to a 1958 B-film thriller, *The Blob*, featuring an amorphous and disgusting carnivorous glob of jelly from outer space that devours people in Downingtown, Pennsylvania. The lyrics of the film's theme song summed it all up: "Beware of the Blob! It creeps, and leaps, and glides and slides across the floor."

Today the movie is part of American folklore, and its imagery in big demand. In July 2016, for instance, Ed Killer, outdoors columnist for *Treasure Coast Newspapers*, invoked it for an article: "'The Blob' Gets Its 15 Minutes of Fame."

This Florida Blob, however, was real, and invading Killer's hometown.

"The microcystis toxic algae showed up in Lake O in the form of a 30-mile-long blob the color of radiator fluid," he writes. "[I]t slid its way 27 miles down the C-44 Canal and through the gates at the locks. By mid-May, The Blob lined the shorelines in the river in Palm City. By Memorial Day, it was settling into downtown Stuart."

Concerned public health officials posted signs along riverbanks, warning people to stay away from the slimy green stuff that kept expanding by the hour. In some places the toxic stuff was inches thick.

Residents complained about a powerful rank smell. Many compared it to the stench of a cow pasture or a sewage pit. Here and there the nasty gunk turned blue, black, or brown. The telltale hydrogen sulfide stench of decaying algae sent people back into their houses, gagging. Health officials warned residents not to touch or ingest the stuff.

The people of Stuart aren't the only ones who've suffered algae attacks over and over again. Blobs also appear where the Caloosahatchee dumps Lake Okeechobee water into the Gulf of Mexico near Ft. Myers. Local resident John Cassani has seen the damage. A former research scientist specializing in aquatic ecology, he now heads the Florida Watershed Council and serves as Caloosahatchee River Keeper. Previously, he spent thirty years in Southwest Florida, carrying out applied research for water restoration projects for various local governments. "When I first arrived in 1978, the Caloosahatchee River estuary had one thousand acres of aquatic grasses," he recalls. "There were massive flocks of waterfowl—tens of thousands of birds—that fed in the estuarine area."

The estuary was once "the grand jewel of that region," Cassani adds, "but by 2000, 95 percent of the birds were gone from the upper estuary as a result of water management decisions and pollution."

Unlike the cinematic Blob, the ones that attack Florida's coasts and so many other places are no alien life forms. Free-floating or "planktonic" algae, in fact, has been part of the food chain almost since life began. Classified as a phytoplankton, it is one of many small organisms that live in water and provide nourishment for other forms of life.

The trouble comes when it is overfed by overdoses of plant nutrients, including nitrogen and phosphorus, released by lawn and agricultural fertilizers, manure, and wastewater. Scientists call the process *eutrophication*—too much of a good thing. Algae goes on a feeding binge when a nutrient banquet appears. An overabundance of certain kinds of algae is deadly. Unlike toxic chemicals and other pollutants, algae reproduces. If left unchecked, it depletes oxygen from the water and wreaks havoc on neighboring aquatic life forms—from shellfish to rooted water plants, including good algae that promotes a healthy water environment. Some algae forms long chains, or filamentous threads, interweaving and producing thick, yucky-looking green mats, which to some people feel like wet steel wool.

What makes the headlines most often are algal blooms, or outbreaks, which take place when algae grows quickly out of control and produces toxins that, if ingested, can be fatal to fish, birds, and manatees. Drinking water with elevated levels of the stuff can also sicken adult humans and maybe kill small children.

Algae also helps heat up the planet, says Peter Barile, a senior scientist with Marine Research and Consulting in Melbourne, Florida. Fly over the Indian River Lagoon, he says, and the big dark spots of algae infestation below tell you why. "Algae creates a black background, which absorbs sunlight, exacerbating global warming and making it harder for us to deal with sea rise."

Worse yet, these ancient microorganisms may be adapting to a host of nutrients humans keep flushing into water bodies. "Algae is evolving to do whatever it wants," Barile explains. "We're dealing with an adaptive organism that can outpace our ability to effectively respond to it."

Researchers now believe some algae varieties generate as many as five hundred new generations every year, making it more probable that strains will keep emerging that will be resilient to whatever humans try to do to eradicate them.

Adding to Florida's misery are murderous red tides. Unlike their freshwater-loving algal cousins, a saltwater variety (*Karenia brevis*) is to blame for most red tides in the Gulf of Mexico and Florida. Though naturally occurring, they now are feasting on the ever-increasing nutrients flowing to the coasts. Rising temperatures may also be helping them thrive.

Appearing together, algal blooms and red tides are extra deadly. This became clear in the summer of 2018, when a red tide exploded in canals and coastlines from Tampa Bay to the Florida Keys, killing thousands of fish, manatees, sea birds, sharks, and even whale sharks. Meanwhile, blob-like freshwater algae was running amok across the state. The situation was so dire the Florida governor declared a state of emergency.

The Costs of Runaway Algae

Algal blooms are costly. According to the Stockholm Environment Institute (SEI) 2012 report, outbreaks of algae and red tide cost Florida between $1.3 billion and $10.5 billion each year.

Tourism is one of the biggest losers when these menaces arrive. Owners of charter fishing boats, bait and tackle shops, motels, marinas, restaurants, and others close down for a long time. And once the emergency subsides, local business owners don't know if they should bother to reinvest, because unless changes are made, another algal attack will come.

Algae's economic wallops can occur quickly. "Water quality of the St. Lucie Estuary, Loxahatchee Estuary, and the portion of the Indian River Lagoon north of the St. Lucie Inlet . . . resulted in an estimated $488 million reduction in Martin County's aggregate property value between May 1, 2013 and September 1, 2013," according to a March 2015 report from Florida Realtors, a trade association.

Fixing the problem is also expensive. It would cost $4.6 billion to reduce nutrient loading in the Indian River Lagoon area to be sustainable again, according to a 2015 estimate from the East Central Florida Regional Planning Council and the Treasure Coast Regional Planning Council.

The Fertilizer King

The fact that algae is so deadly is ironic, given that its modern widespread existence can be traced to a largely forgotten scientist, Fritz Haber, a brilliant German who wanted to save lives. In 1909 he discovered a process that revolutionized the making of a synthetic fertilizer that provides the nitrogen that food crops need to grow.

Haber's importance is easy to see if you visit the website science heroes.com, which ranks world scientists by the estimated number of human lives saved by their contributions. There he is at number one, along with colleague Carl Bosch, the man who commercialized Haber's process, which the webmaster estimates saved 2.72 billion human lives that would otherwise have succumbed to starvation.

More than a century ago, people needed a synthetic fertilizer because there wasn't enough naturally produced nitrogen available in nature to fertilize all the crops needed to feed the growing world human population.

Nitrogen is not scarce, however. In fact, nitrogen gas makes up about 80 percent of the atmosphere, but most plants can't directly absorb nitrogen from the air. Instead, they rely on an intermediate step called "nitrogen fixation." Soil bacteria change nitrogen gas into water-soluble forms that plants absorb into their tissues.

Haber came up with an efficient, relatively cheap, high-temperature method of extracting nitrogen from air and turning it into a compound that could be used in making fertilizers. Soon, factories were using the new process to mass produce it on an industrial scale. Without manmade fertilizer, there would be no way to grow enough food to sustain the more than seven billion individuals who now reside on earth.

Not everything Haber did was so human friendly. He was also the father of a poison gas used by the German army in World War I. Nitrogen fertilizer was an ingredient in German bombs. A derivative of an insecticide Haber developed was used later to make Zyklon B, a gas used by the Nazis for extermination purposes at certain concentration camps. This was a strange twist in events, since Haber was a persecuted Jew.

Nonetheless, in 1918 he received the Nobel Prize in Chemistry for his work in nitrogen fixation. Most likely, Haber never imagined his prize-winning achievement would one day contribute to the runaway growth of a deadly monster.

Not the Only One

Nitrogen, however, has a trouble-making companion, phosphorus—another element all plants and animals need to survive. Many Florida soils are rich with it. In fact, Florida—along with China, Morocco, and part of South America—has the biggest phosphate deposits in the world. So much, in fact, that Florida phosphate is mined in the state and sold internationally as an ingredient for fertilizer.

Together, these two nutrients feed algal blooms that cause mayhem worldwide. According to the United Nations, "Globally, the most prevalent water quality problem is eutrophication, a result of high-nutrient loads (mainly phosphorus and nitrogen), which substantially impairs beneficial uses of water."

In August 2014 toxic algal blooms caused authorities to close down the water supply system in Toledo, Ohio. Satellite photos reveal that Lake Erie at times appears green from space. In 2015, along the U.S. West Coast, sea lions, seals, and porpoises fell ill or died from exposure to a neurotoxin, domoic acid, produced by certain species of marine algae.

Deadly algal blooms are growing in frequency and intensity in the Americas, China, the Mideast, Australia, Africa, and Europe. An estimated four hundred ecological dead zones occur around the world thanks to lethal algal slime, including an area of oxygen-depletion measuring more than 6,000 square miles in North America's Gulf of Mexico.

Closer to home, periodic outbreaks of toxic algae continue to plague not only Lake Okeechobee, the Caloosahatchee River, the St. Lucie River, and the Indian River Lagoon but also a growing number of springs, rivers, canals, lakes, lagoons, bays, and other Florida water bodies.

Algal blooms aren't brand new to Florida. In 1947 they appeared in Lake Apopka for the first time.

Today algae is so common and uncontrolled in Florida that it could very well be the state's worst pollutant. "Look at the list of Florida's impaired waters," says DEP's Janet Llewellyn, "and you'll see that nutrients [which feed algae] head the list of causes."

Gunking of Florida's Springs

Among the hardest hit of Florida's waters are many of its more than one thousand fabled freshwater springs. Once, Floridians could justifiably declare these sparkling, natural wonders the state's crown jewels. Today, that's harder to claim since some of these liquid treasures have dried up. Others are dying or being degraded by explosive growths of nuisance algae and other reasons.

A century ago, however, they were tourist lures, rightly billed as natural wonders of the world. Unlike seepage springs (places where shallow groundwater puddles at the edge of sloping ground), the springs that make Central and North Florida famous are the "artesian" variety. They are points of pressurized groundwater coming to the land surface found in karst areas, where underground limestone aquifers are full of pores and caves that resembles Swiss cheese. Florida's high rainfall supplies abundant water to the land surface, some of which infiltrates into this porous limestone, saturating it. As the water volume in the aquifer increases, it exerts more and more pressure while seeking the path of least resistance through underground cracks and caves in the limestone to openings that lead to the ground surface, namely the artesian springs.

This escaping groundwater forms a spring pool, a spring run, and eventually discharges into a river, a lake, or a coastal estuary. All day and night this water naturally flows out of the springs. The U.S. Geological Survey (USGS) estimates Florida's springs historically produced an average combined discharge of more than 12 billion gallons of fresh water daily.

But this isn't the only thing that makes Florida springs famous. They are also inexpressibly beautiful. If you were lucky enough to go snorkeling in one of them before, say, 1990—that's about when many springs started to show signs of extreme distress—you were apt to have witnessed water so sparklingly clear that you would almost forget you were in water at all. What you probably would have experienced was a magic world of sun-streaked undulating eel grass, rooted in a sugar-white sandy bottom, while hundreds of fish—mullet, bass, bream, catfish—swam by. There was always a good chance of seeing gators, otters, turtles. Sparkling air bubbles may have flitted upward from somewhere deeper below, where scuba divers probed honeycombed limestone caves. Face down, breathing through your rubbery mouthpiece, you shivered in the perpetual 72-degree water, as the Florida sun gloriously warmed your back.

So clear, so pure, were these springs that they once were the go-to places for Hollywood, television, and commercial crews needing transparent water sites for underwater shots. In 1965 the wonder dog Lassie performed at Alexander Springs for a television episode. An underwater scene from the action film *Airport '77* was shot at Wakulla Springs. At Weeki Wachee Springs, actress Ann Blyth donned a fish tail for the making of *Mr. Peabody and the Mermaid.*

The empress of all springs, however, was Silver Springs. This largest of Florida's first-magnitude springs provided dazzlingly clear underwater sets for six Tarzan films, starring Olympian and owner of five gold medals for swimming, Johnny Weissmuller. Segments of the films *The Creature of the Black Lagoon* and *Never Say Never Again,* along with episodes of the *Sea Hunt* television series, and numerous other commercial films were also shot at Florida's premier spring.

Until Silver Springs lost its mojo and the arrival of Disney World, it was once the state's number-one tourist attraction and generated about $61 million a year of economic impact on the local economy in Marion County.

Today, though less robust than they once were, Florida's springs are still moneymakers. Swimmers, canoeists, kayakers, boaters, anglers, snorkelers, and scuba divers still continue dipping into the chilly spring waters. The Florida Department of Economic Opportunity estimates that in 2011 the annual economic impact generated by four of Florida's largest springs—the Ichetucknee, Wakulla Springs, Homosassa Springs, and Volusia Blue Spring—was a combined total of $68.5 million.

The Persistent Springs Scientist

Florida's springs also impart noneconomic values. That is one of the many points made over and over by springs scientist Bob Knight, whose headquarters is the Florida Springs Institute (FSI), a nonprofit research and educational organization dedicated to the protection of Florida's freshwater springs. The FSI operates out of a two-story red-bricked building that once served as the jail and post office in the small North Florida town of High Springs.

The institute is, in part, a memorial to Howard T. Odum, the "Father of Springs Ecology" whose groundbreaking research on how springs function inspired the creation of the organization. Odum's sterling reputation among other scientists and his landmark publications about Florida's Silver Springs stretch across the state of Florida and around the world.

Knight is one of Odum's many scientific protégés. Now an accomplished wetland and environmental scientist and systems ecologist himself, he is founder and director of the institute. Knight is also on a mission to save Florida's springs from the unnecessary impacts of modern, rampant development. It's a Herculean task.

Undaunted, he'll tell anyone who'll listen why Florida's springs are important, how they are being degraded, and what needs to be done to repair the damage.

One of his public talks took place on a cold January day at FSI's North Florida Springs Environmental Center in High Springs before an audience of thirty. On that day, Knight, tall and trim, white-haired and bespectacled, lectured about springs' biology. With humor and passion and a knack for avoiding arcana, he spoke about the variety of fish, manatees, plants, and other biota commonly found in healthy Florida springs. A spring, he said, is a living thing whose parts are interdependent: "It has a steady heartbeat measured as a daily rhythm of dissolved oxygen, and its metabolism can be measured, along with its respiration—just like any other living thing."

That metabolism occurs when a spring is filled with plants that produce oxygen and creatures that consume that oxygen. When carbon dioxide, sunlight, and water combine in the right balance, the springs produce stored energy or food in the form of plant life. At night, or on a cloudy day, dissolved oxygen levels dip, and on a sunny summer day the oxygen concentration goes up, along with increased food or biomass in the spring.

A healthy spring is as productive as a cornfield, too, explains Knight, when you consider how much eel grass and other aquatic flora it generates.

This productivity, however, begins to vanish in a spring when nuisance algae takes over. It could be that the native plants are stressed out from nutrient loading and can't compete with the filamentous algae, which loves the stuff. Other exotic plants, such as hydrilla, also quickly overrun and smother native aquatic plants.

Other factors affecting spring health include the amounts of light and dissolved oxygen, changes in the number of snails that feed on algae, and the presence of pesticides in the spring water.

Knight also used another analogy to make a different point: a perfectly tuned Lamborghini automobile. "And then somebody messes around with the engine until it's sputtering and performing badly. That's what we've done to the springs. They don't and can't function like they did before they lost their flow and water purity."

He became emotional when he talked about the "good algae" that formerly provided most of the productivity in Florida's healthy springs, and the "bad" blue-green algae with its long, filamentous strands that cover the native plants and formerly white sands. Bad algae now dominates aquatic environments throughout the Crystal River springs and Kings Bay, Silver Springs, Rainbow Springs, additional springs that feed the Santa Fe River, and many other degraded water bodies. In many of the altered springs, some or all of the native plants, such as eel grass, are now gone, smothered by the nuisance algae that covers their leaves. Lots of fish have disappeared as well. Knight and other scientists have documented a more than 90 percent decline of fish biomass at Silver Springs alone in the past fifty years.

"Nothing wants to eat this filamentous algae—it's not palatable." he said. "Aesthetically, the noxious algae looks disgusting for anyone who's lived long enough to have seen the changes at the springs."

In his book *Silenced Springs*, Knight reports that nitrate concentrations in the Floridan Aquifer are anywhere from fifty to one thousand times higher than their natural background level in the groundwater and are "at levels that can lead to explosive algal blooms in springs and rivers, and approaching levels that are acutely toxic to human infants."

Adults should be wary too. According to the federal government's Centers for Disease Control, there is "limited evidence that nitrite [a form of nitrate] may cause some cancers of the gastrointestinal tract in humans and mice." Their research has found that certain groups of humans consuming water with as little as two parts per million of nitrate have a threefold increased risk of various cancers and birth defects.

Elevated nitrate in drinking water derived from the Floridan Aquifer should be of concern to all of those living in the karst areas of North and Central Florida. However, those people living closest to the biggest nitrogen polluters—namely, confined animal feeding operations (dairies and chicken houses), agricultural row crops, golf courses, and urban areas with high septic tank densities—ought to take note that higher than average nitrate levels are commonly found in private wells.

In fact, wells in the most highly developed agricultural areas in Gilchrist, Suwannee, and Lafayette Counties, and older urban areas dominated by septic systems and private wells, have shown concentrations higher than thirty parts per million and some above one hundred parts per million, ten times the EPA safe drinking water limit.

Later that day at a nearby eatery, Knight mentioned his favorite reading material during his early college days: *The Lord of the Rings* trilogy and its main theme: "The powers of darkness against the power of light and goodness."

Is that how he sees the battle for Florida's waters? The question made him blink. He pauses. "Yes," he answered. "That's what's going on in Florida."

To Knight "all groups and individuals who profit from extracting excessive groundwater and who avoid taking responsibility for curbing their own pollution, in the name of higher profits, are guilty of the decline of Florida's springs."

Springs' Tough Guy

Florida's springs are also in danger of being loved to death. For much of the twentieth century, several of Florida's more spectacular ones were privately owned tourist attractions that brought hordes of visitors from around the world. Many springs are still in private hands. City and county governments also own some. Others, such as Juniper, Alexander, and Salt Springs, are under federal control in the Ocala National Forest, and because these springs are located farther from human settlements, are in some ways still reasonably pristine.

Today, many of the most famous springs—Wakulla, Weeki Wachee, Wekiwa, Silver, De Leon, Manatee, Rainbow, Homosassa, and Volusia Blue—have been acquired by the state and turned into state parks. "They are the best of the best," says Tallahassee's Jim Stevenson, formerly a senior biologist for the Florida Park Service. He's also been a tough, outspoken Florida springs defender for more than forty years. His deep, ringing voice has commanded attention from rapt audiences across the state.

Florida springs' show-biz days, however, Stevenson says, are mostly gone, because they had to compete with high-tech theme parks and water problems. "Today newcomers to Florida visit the springs and think they're gorgeous, but they're highly degraded. They're sacred to everybody—until it's time to do something to help them."

By this Stevenson means the homeowners who balk at being restricted on the use of fertilizer for their "perfect" lawns. He also criticizes the big "influential opponents" of springs, which, according to Stevenson, are major lobbying groups for Florida industry and agriculture, along with realtors, builders, and municipalities when they find they'll have "to pay for some of the solutions for the problems they have caused."

Few Floridians have done as much as Stevenson to fight for springs. For decades, he's been an outspoken champion for them and spearheaded numerous springs-protection measures, conferences, and task forces. In 1999 he took Governor Jeb Bush and DEP secretary David Struhs on a canoe trip on North Florida's Ichetucknee Springs, so they could see for themselves that the state's beloved springs were in trouble.

The trip worked. Struhs authorized Stevenson to create a Springs Task Force, whose mission was to assess the status of Florida's springs and what state government should do to help. Bush followed up during the 2001 legislative session with the Florida Springs Initiative, which provided $2.5 million annually for springs protection. It was, says Stevenson, "the first comprehensive springs protection program in the world."

Before becoming a naturalist, Stevenson worked as a land surveyor and caught snakes and alligators with his bare hands and then sold them to Gatorland, a tourist attraction near Kissimmee. Though now retired, he's still a major force in Florida environmental battles. "Don't mess with our springs!" is his motto.

These days Stevenson leads tours from Tallahassee to Wakulla Springs, formerly the iconic landmark meeting place for powerful politicos, located near the state capital, to drum up public support for the ailing spring.

Nobody today photographs very many underwater scenes for Hollywood or television at Wakulla Springs, Silver Springs, or many other Florida springs anymore. Fish and other aquatic creatures are nearly absent. Native underwater plant life is coated with green snot-looking algae. Dazzling clarity is a memory.

Much of the commercial theme-park kitsch has also vanished at many of the springs, except maybe at Weeki Wachee, where you can still see beautiful mermaids, all of them state employees, cavorting underwater at depths down to ninety feet.

Making Matters Worse

It's bad enough that bad algae and other pollutants make the springs unhealthy. But even worse, many are also being overpumped.

Think of a nineteenth-century doctor bleeding an already ailing patient, and you sense the problem.

Knight and other scientists believe data they've acquired from their own studies along with a slew of technical reports published by state scientists prove that too much extraction of groundwater from the aquifer by too many wells is reducing the pumping power of Florida's springs. It's an assertion that sometimes puts them at odds with state policymakers.

Some of the formerly large, recreational springs—like Kissingen Springs in Polk County, Worthington Springs on the Upper Santa Fe River, and White Springs on the Suwannee River—don't flow under most conditions due to increased groundwater pumping.

Other springs are having trouble, too. "Discharge data from 393 of the state's 1,000+ artesian springs were used to estimate trends in total spring discharge by decade since 1930–39," according to a February 2016 article coauthored by Knight in the *Journal of Earth Science and Engineering*. "This analysis indicates that average spring flows have declined by about 32% . . . Existing groundwater pumping rates from the Floridan Aquifer in 2010 were more than 30% of average annual aquifer recharge, and allocated groundwater use in north-central Florida is nearly double current estimated uses."

The specific figures for the more famous springs are discouraging, too: "Silver has lost about 30 to 40 percent of its average flow," Knight says. "Rainbow about 20 percent, Ichetucknee about 25 percent, but Wakulla's flow is rising as it pirates groundwater originally headed to Springs Creek, its coastal neighbor."

Decreased flow also causes other problems. When spring flow is reduced, the bad, filamentous algae remains in place and grows over everything else. Worse yet, there's also less water to dilute the toxic water.

However, as new studies show, if flow is increased, the nasty stuff has trouble colonizing on aquatic grasses. Even algae already growing can be flushed downstream if flow is fast enough.

Lack of rain is one cause of reduced flow. But Knight and other scientists have little doubt overpumping by humans is largely to blame, and they insist that the problem grows worse every day as the water districts keep on issuing new consumptive water permits. The aquifer, they explain, is like a hose with water flowing through it. The end of the hose where the water comes out is analogous to a spring. Poke thousands of holes in the hose, and water also shoots off in thousands of different directions. The pressure at the end of the hose can one day become a dribble.

Owning Up to the Problem

None of this litany of woe means Florida's springs have to be doomed. But it does indicate they need urgent and effective protection. There are at least hopeful words coming from Tallahassee. DEP, for example, announced goals of reducing nitrogen by 79 percent at Silver Springs and 82 percent for Rainbow over a fifteen-year period.

"It may seem like a long time, but it's fairly logical if you think about it," says Casey Fitzgerald, director of springs protection at St. Johns River WMD. "Typically, ecologically challenged natural resources have been subjected to many decades of pollution loading. Overcoming these legacy loads and changing human behavior to curtail standard practices that have caused excessive loading will not happen overnight. It also takes time for governments and utilities to determine the most cost-effective projects and to design, permit, budget, and execute these projects."

In July 2016 the Florida legislature produced a new water law, which recognized the state's springs were in trouble and authorized corrective measures. As a result, "DEP has formally assessed water quality for all thirty Outstanding Florida Springs (OFS), and has verified twenty-four as impaired for nitrate," says the department's spokesperson, Dee Ann Miller. For sixteen of these, she adds, the department has already adopted water quality restoration targets. By July 2018 the remaining eight springs will also be covered by recovery plans.

Knight and other scientists, however, are skeptical. They point out that Florida already has laws to curb excessive pollution and depletion of the springs that have been on the books for more than forty years but haven't been adequately enforced. These existing laws and the new one, argue the scientists, will need an added commitment by the state government to actually save any springs.

There are other critics, too. In June 2017 the editorial board of the *Ocala Star Banner* took the water districts to task for accepting "Shockingly inadequate minimum flows and levels for Silver Springs, Rainbow Springs and the Crystal River springs group. These levels will allow already diminished springs to be further degraded due to reduced flows and nitrate pollution fueling algae growth within them."

The newspaper editors also noted that they "remain confounded that the science of the Florida Springs Institute, and other springs advocates, was virtually ignored by the water districts, who repeatedly said their [finding] to lower the spring flows even more 'was based on the best available science,' when one look at the springs tells a tale of degradation and slow death."

When challenged like this, the water management districts' governing boards typically respond they did their duty by approving recommendations prepared by staff scientists. The scientists say they're using solid science based on sophisticated computer models to make their assessments.

Something profound may be overlooked by a public that thinks debates over Florida's springs are limited to how the water bodies look at the surface. They are not just beautiful curiosities. Florida's springs are also windows into the aquifer. What flows to the surface is symptomatic of what lies beneath.

"Cleaning up an aquifer isn't like cleaning up a lake. Once it's polluted, it stays polluted," says Jim Stevenson.

How did Florida's springs and all other freshwater bodies get so messed up? The answer lies in the not-so-distant history of how the state of Florida was conceived and developed—and how it was butchered and bled nearly to death.

4

Drain Me a River

"FLORIDA HAS THE Second Fewest Native Residents of Any State," proclaimed the *Miami New Times* in August 2014. According to the newspaper, a *New York Times* study revealed only 36 percent of residents—roughly one of three Floridians—were born in the state. Moreover, 23 percent were born outside the United States.

If two-thirds of Florida's population has, at best, shallow roots in the state, it should be unsurprising if many residents also lack a deep understanding of the state's history or its pressing problems—like those concerning water. For that matter, how many state leaders themselves have deep memories of Florida's water wars, hurricanes, and tumultuous environmental struggles? After all, as of 2017 the last Florida governor to have held office and be born in the state was Buddy MacKay in 1999.

Florida history isn't even a required high school course in the state's public schools, nor is much being taught about Florida's water problems.

This lack of awareness is a problem because to understand our current water troubles requires a knowledge of the past, suggests State Senator David Simmons, a Republican from Longwood, north of Orlando. Too many Floridians, says the Tennessee-born attorney, don't know their state history, or forget what they were taught—but they should know.

"The way we got where we are with our water problems," he says, "started 150 years ago. Today, people forget that history. They forget that engineers in the past thought the Floridan Aquifer was inexhaustible. Maybe that belief had merit at the time. But what you could once do using an inexhaustible aquifer is not what you can do today with twenty million people."

The backstory is even older.

Origins

The landmass everyone knows as Florida has spent most of its time under seawater. These periods of submergence coincided with ancient global warming events that raised sea levels and swamped the world's low-lying, flat places. Untold billions of sea creatures lived and died and deposited their shells and other remains on the shallow bottom. In time a vast, compressed pile of fragile limestone rock combined with dolomite to provide the underground foundation that lies below the land surface of much of modern Florida. This huge rocky system makes up the aquifer, which is kind of a rocky sponge filled with crevices, cracks, and openings containing vast amounts of water.

It was when the world turned icy that Florida emerged from the sea. Rather, part of it did. That distinctive well-known peninsula is actually the upper level of a bigger geological formation called the Florida Plateau, most of which, to this day, remains below the surface. Thus, Florida's coastlines have been and are always in flux, expanding and retreating in response to how much ice is being stored or melted at the poles and high elevations.

The first humans in Florida, the Paleoindians, who arrived on the peninsula, presumably following mammoths and bison, about fourteen thousand years ago, found a land that was much drier and cooler than it is today. An ice age was still chilling the planet and lowering sea levels by stockpiling water as glaciers in the mountains and ice sheets at the poles. This meant Florida's underground fresh water table was much lower than it is today, and its landmass perhaps twice as big. The Paleoindians naturally often lived near springs and rivers. To visit some of the remains of those settlements today, archeologists must travel by boat several miles offshore, don scuba gear, and conduct their archeological examinations along the sandy bottoms of the sea.

The climate had changed drastically by the time Spanish explorers arrived in Florida during the sixteenth century. These adventurers sought fortune and bore muskets, germs, and a culture totally alien to that of the indigenous peoples.

Florida had also become wetter by this time. The world had been slowly warming from around 9000 BC, which meant the Spanish experienced a climate similar to that of today. Sea levels had risen. Florida's coastlines had retreated. Swamps, wetlands, estuaries, lakes, ponds, and rivers were now more plentiful. A third of a million indigenous inhabitants may have been living along the coasts of these water bodies when the conquistadores arrived.

Fresh water confronted the Spaniards everywhere. Rivers, lakes, and swamps, along with immense glassy sheets of water, extended from one horizon to the other. Even the air itself was thick with moisture. Mighty, crystal-clear springs bubbled by the hundreds among the rolling hills. Rising and falling sea levels left behind a legacy of estuaries, bays, inlets, and wetlands.

It rained a lot, too, especially in South Florida. And in late summer, dangerous tropical storms and hurricanes whooshed in, flooding the peninsula with even more water.

The Spaniards left few lasting traces on the Florida landscape, and they didn't tamper much with the hydrology. Neither did the French nor the British, when they took their turns occupying the land.

Americans Make Their Mark

The Americans were different. By the end of the Civil War in 1865, Florida counted itself among the defeated and dispirited former Confederate states. The state was poor and backward. It was also thinly populated. Only about 140,000 people lived in the state, and most of their homesteads were located close to the Alabama and Georgia lines along an east-west stretch of land from Pensacola to Jacksonville.

Much of the state was buggy, humid, and soggy. Nobody controlled mosquitoes, or yet understood that these insects often carried diseases. Many people believed the stench of bad air (*mal-aria* from the Italian) that came from places such as swamps and stagnant pools was what made people sick. Small wonder many Floridians shut their windows to keep out the foul air, which, of course, kept out disease-bearing mosquitoes, as well.

There were plenty of wild beasts—as is still the case—with sharp teeth lurking about, too: bull sharks in the lagoons, alligators in all the freshwater bodies—and crocodiles in the coastal, brackish, and saltwater areas of South Florida—along with panthers and black bears almost everywhere. Venomous snakes—rattlers, cottonmouths, and coral—were also abundant.

Along with these off-putting realities, Florida also had its alluring qualities. For one thing, much of the state was graced with a primeval, subtropical beauty. And it wasn't really hot all the time. In fact, during winter temperatures dropped below freezing in part of the state, and sometimes snow fell along the Pensacola to Jacksonville corridor.

Best of all, there was plenty of vacant, cheap land. Why wouldn't there be? After all, Florida's indigenous peoples who had lived on the peninsula for thousands of years—the Calusa, Ocali, Timucuan, Alachua, and others—were now mostly gone, having succumbed to European-introduced diseases, wars, and cultural displacement.

Nearly absent was Florida's best-known tribe, the Seminoles. Unlike the other native peoples of North America, these one-time Creeks from Georgia and the Carolinas had been living in Florida only since the 1700s. By the mid-nineteenth century, federal troops had forcibly relocated a majority of them to reservation lands in Oklahoma to make way for white settlement. Only a few hundred defiant, belligerent souls remained behind, hiding from federal troops in the almost impenetrable Everglades.

Despite an abundance of unoccupied land, the state's power barons wanted more. This goal could be accomplished, they argued, if Floridians would just wage a new war, this time not against Yankees or Indians, but instead against a ubiquitous foe: fresh water.

"Florida is so watery and vine-tied that pathless wanderings are not easily possible in any direction," wrote John Muir, a Scottish-born, bearded, peripatetic naturalist whose famous one-thousand-mile hiking trek across America took him into Florida's wild heartland during the burnt-out wake of the Civil War.

Muir, however, may have had more luck if he'd used a boat. "So omnipresent and multiform was the water that you could get in a canoe on the Georgia border and except for some portages, paddle the length of Florida," writes author T. D. Allman. "You would traverse maybe seven hundred miles of waterways called by different names 'lakes,' 'rivers,' 'swamps.' Whatever you called them, the pressure of Florida's immense quantities of fresh water kept you afloat."

In 1877 Jim Buck, for instance, witnessed that wet plentitude. He was a young writer from Cambridge, Massachusetts, yearning for a bit of travel and adventure, so he moved to Coconut Grove on Biscayne Bay, a few miles south of present-day Miami. There he built a shack and planted a garden.

He also amassed a trove of personally written impressions of early Miami. Among them was this account of how easy it was to tap a local water supply: "Springs of good water are common and wells are to be had by a comparatively small amount of digging. This latter may often be done with axes, through the soft rock, so that the well is already stoned, when the water is reached after a few feet of cutting. Many springs burst up through the bottom of the bay, and we see fresh water boiling up through the salt."

At the time, the nearby Miami River was pure and clear. A section of it was rapids. There was even a waterfall. Buck further reports seeing the river "bursting, as though propelled from a hydrant, out of the vast . . . accumulation reservoir of the Everglades, and forcing its widening way to the Bay."

Today, no sane Floridian would dip a cup into the Miami River for a drink. For a long time, it's been a filthy, rank, dark, urban waterway. In the 1990s, two federal grand juries declared it a cesspool. The Seybold Canal/Wagner Creek, a tributary of the Miami River, ranks as one of Florida's most polluted waterways.

Environmentalists aren't the only ones who worry about the health of the Miami River. Having run out of seafront space to erect ever more condos, developers, too, want a rehabilitated river to make it attractive-looking enough to risk building a Riverwalk, replete with sixty-story condos with penthouses, restaurants, retail stores, and so forth. As a result of these concerns, an estimated $20-million river-cleanup project, involving dredging and excavating contaminated muck, got under way in 2017.

Natural Discharges

A century or so ago, most Floridians would have questioned the sanity of anybody wanting to spend money restoring a river. Instead, their big water problem, as they saw it, was there was too much of it, clean or dirty. And they wanted to get rid of as much of it as they could.

Nature had provided a natural drainage system for Florida's huge mass of water, though few of the state's early residents appreciated it. In general, the terrain in North and Central Florida is higher than in South Florida; Florida's interior is generally higher than the coasts. Therefore, water drainage tends to flow southward and out to sea.

The state's big rivers evacuated much of Florida's huge water supply. In North Florida, the St. Mary's and the St. Johns Rivers emptied into the Atlantic, while the Apalachicola and the Suwannee flowed west into the Gulf of Mexico.

In Central and South Florida, drainage was a more spread-out process. The Kissimmee River bled into swamps in the state's interior and traveled south into the 730-square-mile Lake Okeechobee—the state's largest freshwater lake. When the summer rainy season came, Lake O—an affectionate nickname—often overflowed its banks, swamping the surrounding region. This sheet flow was wide and shallow, about one hundred miles long and sixty miles wide. It was part of an even bigger drainage area that extends over two hundred miles.

This was the Everglades.

Before humans tampered with it, this slow-moving river took up four thousand square miles and occupied about 33 percent of the Florida peninsula. It was so big that some of the earliest explorers thought they were paddling in a watery world amid unconnected islands. They were wrong. The reality was that they were floating inches above a landmass inundated with countless wetlands, marshes, bogs, and swamps, dotted with islands of pines and hammocks. The huge, sluggish waterway crept ever southward and slightly westward as it followed the slope of the land at a speed of a mile an hour. At the end of its journey, it emptied into Florida Bay, the biggest estuary in North America, located on the state's extreme southwest coast.

A Time to Drain

During the second half of the nineteenth century, few Floridians appreciated the Everglades as a majestic work of nature. Instead, popular opinion held that this huge swamp—just like all other swamps, marshes, and bogs—was a nuisance and an obstacle to human progress.

Such thinking was in step with a belief in Manifest Destiny—the strong conviction held by Americans that they had a duty to expand their nation's borders and the American way of life. Nature was something to be conquered, subdued, and put to useful purpose.

So, while other Americans were "taming the West," Floridians were determined to "reclaim" land that lay smothered by fresh water. Hopes ran high that draining much of this water to the sea would entice settlers and business people to the state. Farmers, especially, were expected to put the exposed earth to productive use, as God had intended.

Drainage also was expected to reduce the deadly and destructive scourge of flooding. Many Floridians hoped dredging channels and shaping rivers would also improve transportation for steamboats and small craft, making it easier to travel from one remote riverside settlement to the next.

Getting rid of water for development was hardly a novel idea. Human beings had been doing it around the globe for a long time. "Some 60% of wetlands worldwide—and up to 90% in Europe—have been destroyed in the past 100 years, principally due to drainage for agriculture but also through pollution, dams, canals, groundwater pumping, urban development and peat extraction," writes noted physicist and climate expert Joe Romm.

In the 1700s, America's early European settlers were turning wetlands into farmlands, especially in the Southern colonies, such as North and South Carolina, Virginia, and Georgia. They drained even more during the first half of the nineteenth century, when the United States purchased the Louisiana Territory, annexed Texas, and took land formerly belonging to Mexico, as a booty of war, and opened up the new lands to settlement. The Lake Erie marshes of Michigan and Ohio's Black Swamp disappeared altogether. Many wetlands also vanished in Minnesota, Iowa, North and South Dakota, and the Central Valley of California.

There was plenty to drain in Florida, too. So much, in fact, that in 1842, the U.S. Congress resolved that the Secretary of War should send men into the Florida wilds to report on the "practicability and probable expense of draining the Everglades of Florida."

The man picked by U.S. president James Polk to lead the expedition was a Harvard-educated, slave-owning, Jacksonville citrus grower and attorney named Buckingham Smith. For five weeks during the summer of 1847, Smith led a survey team through the Everglades. A year later, many of the state's most powerful men were poring over his final report to the Thirtieth Congress, in which he extolled the great swamp's beauty in poetic detail.

Of greater interest, however, was what he wrote about the challenge and opportunity offered by a drainage crusade: "The Everglades are suitable only for the haunt of noxious vermin, or the resort of pestilent reptiles. The statesman whose exertions shall cause the millions of acres they contain, now worse than worthless, to teem with the products of agricultural industry . . . will have created a state."

Smith thought draining the Everglades would free up land for cattle and crop production. He also recommended deepening and channelizing the natural waterways that were already draining Lake Okeechobee to allow even more water to flow to the sea. Smith also advised cutting new water paths through the Atlantic Coastal Ridge, a slightly elevated, five-mile-wide strip of limestone that runs along Florida's southeast coast and forms a natural barrier to water drainage from the interior of the state. It also provides the eastern rim of the Everglades.

In 1850 Senator David Levy Yulee, a prominent Florida politician and the nation's first Jewish U.S. senator, led a multistate campaign to pressure the federal government to hand over more than sixty-five million acres of the nation's wetlands (though they weren't called that yet) to the individual states. In return, the states promised to "improve" the properties. That meant, of course, to drain them.

Florida's share of the swamp giveaway was twenty million acres. To speed up the drainage process, state lawmakers passed a law authorizing Florida to give away more than three thousand acres of the newly acquired federal properties, along with other incentives, for every mile of railroad track that anyone laid down. Yulee, who owned a railroad, was one of the beneficiaries.

Few, if any, at the time could imagine the future negative consequences on water resources from this real estate giveaway. By 1956, however, two federal government biologists, Samuel Shaw and C. Gordon

Fredine, had determined that almost all of the wetlands given to the states a century before had ended up in private hands.

"The landowners can do with them as they wish," they wrote. "It is unfortunate that water-conservation and waterfowl protection areas were not selected and set aside for public benefit at numerous locations before the lands were transferred from federal ownership. If this had been the case, the government would not now be in the position of buying these 'wastelands' at high prices."

In the nineteenth century, however, state leaders were doing everything they could to encourage Floridians to drain and develop. It soon became clear that their incentives were responsible for only a handful of small-scale projects that made little difference. It was time, everyone realized, for an undertaking on a grander scale.

A Millionaire Takes Charge

The man who seemed capable, wealthy, and eager enough to carry out this task was the Philadelphia sawblade manufacturer Hamilton Disston. During a fishing trip to Florida in 1881, the thirty-six-year-old mustached industrialist struck a business deal with then Florida governor William Bloxham. For his part, Disston agreed to drain up to twelve million acres of soggy land in Central and South Florida that flooded every time the Kissimmee River Basin and Lake Okeechobee overran their banks. In return, he could keep half of all land that he "reclaimed."

The state, however, soon ran into trouble upholding its end of the bargain when its creditors demanded a sell-off of public lands to raise cash to pay the government's debts. One million dollars was at stake.

So the daring young entrepreneur from Philadelphia again bargained with Tallahassee officials. This time he agreed to buy four million acres of swampland at twenty-five cents each, which he intended to drain and develop. Thus, with a stroke of a pen, Disston helped Florida avoid financial ruin and at the same time may have pulled off the largest land purchase by any one person in history. In fact, he now owned half the state.

By 1882, Disston had two huge dredging machines at work. His goal was to dig a drainage path from the Kissimmee River floodplain to Lake Okeechobee that would connect to the Caloosahatchee River west of the lake in order to haul excess water to the gulf. Disston also wanted to canalize the St. Lucie River on the lake's east side to drain lake water to the Atlantic but couldn't do so because he lacked funds. In 1916, however, others began this work.

Railroads Arrive

As Disston endeavored to dry up Florida wetlands, two other business tycoons were busy putting down railroad tracks into the state. These men, too, believed there were big untapped economic opportunities in the swampy, but sunny and warm, Sunshine State.

Henry Flagler was one of them. Born into poverty in New York, he was a poster child for the rags-to-riches myth popular at the time. Hardworking, shrewd, and ruthless, Flagler became the business sidekick of the world's first billionaire, John D. Rockefeller, and the "brains" behind the Standard Oil Company. Now rich and experienced, Flagler owned his own company that was busily laying tracks and building luxury hotels along Florida's east coast.

Meanwhile, Henry Plant, a Connecticut shipping magnate and developer, was expanding a rail system that ultimately traveled along Georgia's Atlantic coast, and then veered through the center of Florida to Tampa, where he built his own hotel.

These two undertakings marked the beginning of a railroad building spree that would eventually crisscross the state with tracks, providing new transportation possibilities for tourists and settlers alike.

Soon, Americans streamed into the state. Some came by rail and settled in one of the new communities that popped up along the new railroad tracks. Others arrived by steamboat or mule-drawn wagons. They bought reclaimed swampland, built houses, planted sugar cane and vegetables, and established cattle herds.

Meanwhile, steamboats were churning down rivers, carrying citrus, lumber, turpentine, and other goods from Florida's interior to the new coastal cities growing larger every day.

Hamilton Disston's company not only drained the swamps; it also tried to woo customers from as far away as Europe with real estate offers for as little as five dollars an acre. For a while, it seemed as if the Ohio businessman just might earn a place in the gallery of business titans of the era.

But in 1887, Disston ran into financial trouble when a state commission reviewed his work, found some of it successful, but insisted that drought, not his draining efforts, was responsible for most of the dried-up land north of Lake Okeechobee. The commission even recommended that Disston be denied the full acreage he'd been promised.

Other critics piled on, some claiming Disston's labors had little effect on Lake Okeechobee, which overflowed when heavy rains came. These charges may have been unfair. It was more likely, say some historians, that flooding occurred *because* too much water was draining into Lake Okeechobee and overwhelming the evacuation capacity of the Caloosahatchee. "The lake, like a trillion-gallon tub with a tiny clogged drain, continued to rise and overflow in the Everglades in summer storms," explains Michael Grunwald, author of *The Swamp: The Everglades, Florida, and the Politics of Paradise.*

The criticism faded, but as Florida historian Charlton W. Tebeau points out, despite Disston's best efforts, it soon became evident that "drainage could not be accomplished piecemeal in the vast related areas and the cost was beyond the possibility of private undertaking."

Misfortune kept hurling itself at Disston. The Panic of 1893—a serious, nationwide economic depression—took a toll on his already stressed financial holdings. A year later the federal government canceled the tariff on imported sugar, which hurt Florida sugar growers. In 1895 two freezes ruined crops in Florida. For the moment more farmland wasn't needed. To stay solvent in the wake of all these economic blows, Disston mortgaged his assets.

In April 1896 his life took a grim turn. One night he was in Philadelphia, dining with the city's mayor. After the meal, Disston retired to his hotel room and in the morning was found dead. According to popular news accounts at the time, Florida's biggest landowner had shot himself to end his financial despair. This seems unlikely, however, since most of his obituaries reported he was a victim of heart disease. Moreover, as Grunwald points out, Disston left behind the equivalent of $2 million in assets, an impressive sum that suggests he was far from bankruptcy.

After Disston's death, the Florida state government wanted a portion of his other assets and snatched back some of his lands in lieu of back taxes and sold the rest at very low prices.

A New Water Messiah

Florida's great drainage dream, however, did not die with Disston. In 1901 it was kept on life support when state lawmakers passed the Drainage of Counties Act, which provided Florida landowners the right to drain their own properties. During that same year, the federal government gave Florida ownership of the Everglades.

A new water-draining messiah appeared. He was a burly, walrus-mustached, former riverboat pilot, sheriff, and gun runner with an unforgettable name who became governor of Florida in 1904. "My way-back cousin . . . was Governor Napoleon Bonaparte Broward," writes journalist, author, and Florida State University English professor Diane Roberts, "[who] wanted to drain the Everglades, make some money off all that land doing nothing but lying around underwater."

Three years later, with the newly created Everglades Drainage District at his disposal, Broward got to work. His ambitious plans, which called for Lake Okeechobee to be drained with six canals, weren't all business related. This became clear when he publicly insisted that the U.S. Army should round up all black people in the state, transfer them to a distant place at taxpayers' expense, and use military force to keep them from reentering Florida.

The governor's racist plan never materialized. It was also odd, since at the time many African Americans were escaping Jim Crow Florida and heading north of their own free will.

But his views are not forgotten. In 2017 Broward's statue was removed from the Broward County Courthouse, the result of a protest campaign.

Despite his faults, Broward showed that he also had the grit and ego for big-scale projects, despite the ridicule of howling critics who charged the governor with pursuing a costly boondoggle. By 1929 the drainage district he helped to create had carved 440 new miles of canals and levees into the face of Florida. This champion of drainage, however, never lived to see the final outcome, because he died in October 1910, shortly after having won a race for the U.S. Senate.

South Florida Real Estate Bust

By the early 1920s, Florida real estate was hot, thanks in large part to all that available new dry land. It was now the Roaring Twenties—the

Jazz Age. Prohibition was the law of the land, and it was violated day and night. Wild young women with bobbed hair and short skirts were smoking and drinking and riding off in their boyfriends' Model T's far away from chaperones. Evolution was being taught in public schools. Motion pictures depicted men and women kissing. Florida's fundamentalist pastors proclaimed Judgment Day was close at hand.

America's prosperity was more fizz than substance, however, as many economists later reported with the clarity of hindsight. Those few who did see the economic dangers at the time were largely dismissed as party-poopers. After all, the Great War was over. The stock market was booming, and credit was easy. America was awash with tales of easy takings on Wall Street.

Get-rich stories were gushing out of Florida, too. Miami realtors reportedly were becoming millionaires overnight.

With dreams of gaining easy wealth, tens of thousands of Americans rushed into the state. Many risked everything, eager to buy a chunk of their place in the sun even if the land they'd purchased was sight unseen. A carnival atmosphere emerged in South Florida, as giddy and gullible northerners kept showing up at the rail station in Miami—the epicenter of real estate mania—where they were besieged by glad-handing salesmen offering the chance to buy plots of land with inconceivably low down payments. Chances were high that buyers might never see that tract they bought on the street, because they could easily resell it at a profit in just a few hours.

Day by day, land prices soared as the real estate bubble grew larger. The *Miami Herald* was thick with advertisements for real estate, homes, and almost anything else—the most ads of any other newspaper in the nation. One particular daily issue of the *Miami News* had 504 pages.

Traveling to South Florida was now a simple matter. A train or a bus could take you to the Promised Land. Or, better yet, you could drive there yourself in one of Henry Ford's "Tin Lizzie" automobiles. Tourist campgrounds cluttered the roadsides. One huge congress of tent-toting winter tourists in Tampa called themselves the "Tin Can Tourists of the World."

Florida lawmakers tried to speed up the trip for the incomers by raising the speed limit to forty-five miles per hour—ten miles faster than the national average. Car drivers didn't even need a license. Motorists, however, had to be careful because Florida was still an open-range state, and cows roamed freely across highways.

There were other enticements as well. Florida law let you keep much of your riches, because the state had no state income tax. And no inheritance tax!

By the mid-1920s South Florida's real estate craze had grown wilder. Demand for building materials and almost everything else skyrocketed. One train pulled into the Miami railyard with two thousand freight cars filled with cargo. Authorities had to impose an embargo on future deliveries, as they struggled to find enough workers to unload the shipments and a place to store them.

In four years, the number of paved streets in Miami increased from 32 to 420. Assessed property values jumped 560 percent. In November 1924 total deposits in Miami banks were $37 million. By the next year they'd jumped to $210 million.

Miami's population went up too, topping one hundred thousand. Meanwhile, the state was home to one million people, as even more rushed in. Land prices kept going up, as did the number of new homes, stores, office buildings, and casinos. Churches were being built, too, although congregations often tore them down and sold the land when land prices went up.

Jealous and worried bankers and business leaders in Ohio, California, and elsewhere in the nation resented the outflow of cash and warm bodies from their localities and launched a public relations smear campaign to raise doubts about Florida's real estate market.

Some developers, meanwhile, tried to lure buyers from the North by giving their properties classic names such as Venice and Naples. Mecca Gardens was created in Ft. Myers, while Opa-locka was marketed as the new Baghdad. Developer George Merrick created the City of Coral Gables, replete with patios, arches, and plazas to impart a Mediterranean look. His promise that a university was being planned sent a signal to the public that all the flurry of land sales and the runaway construction were not frivolous, but lasting contributions to a permanent and important new city.

One of the most ambitious men among this crop of developers was "The King of Dredge and Fill Development"—Carl Fisher, an audacious entrepreneur and builder of the Indianapolis Speedway. In 1912 Fisher moved to South Florida, bought a mangrove swamp located a few miles off the coastline of downtown Miami, and set out to build a man-made paradise. His workers razed mangroves and wiped out the habitat of Florida crocodiles and numerous birds. They scooped up enormous amounts of sand from the bottom of Biscayne Bay and dumped it all over the mutilated swamp until an island was made. Upon this man-made, barren sand-pile rose the condos, upscale homes, marinas, restaurants, clubs, and hotels today known as Miami Beach, which today is being slowly flooded by sea rise.

All of this building frenzy, however, couldn't go on forever. In early 1925 the Florida real estate market faltered. Complaints of fraud were mounting. Sensing the party was over, the Northern press now began publishing articles warning of the folly of investing in Florida real estate. Bankers, at long last, sobered up and wondered if land prices were too high.

As demand for subtropical land fizzled out, many South Floridians became jittery and dumped their investments. Sales began to slide. Soon, there was a glut of property, and by 1926 the bubble had burst, sending the state into economic decline.

Another financial catastrophe, meanwhile, was on the way that would sink the entire nation and much of the world. But first Florida was to experience something even worse—something deadly.

5

Replumbing the Great Florida Outdoors

DURING THE SUMMER OF 1926, a nameless hurricane hit South Florida and plowed into the Everglades. Most Floridians who were new to the state had never imagined such a destructive storm was possible and were caught off guard when it struck.

During the tropical onslaught, a crudely built earthen dike erected by local settlers to confine the southern perimeter of Lake Okeechobee collapsed. Floodwaters swept through the pancake-flat countryside, destroying wooden homes and buildings from the Everglades to the Atlantic coast and south to Hollywood, Hallandale, and Miami. By the time the counting was over, the storm had killed four hundred people and left forty thousand homeless.

Two years later another killer storm hit. This one was so powerful that it blew water out of Lake Okeechobee and drowned more than twenty-five hundred people in Moore Haven and other nearby small towns. For several days horrified rescuers gathered up the bloated, decomposing bodies in the flooded sugar cane fields. The ground was too wet to dig proper graves, so rescue workers placed the remains of white victims in coffins and carried them elsewhere for burial. Bodies of African Americans, though, were interred in a mass grave in West Palm Beach. Hundreds of decomposing corpses, too far gone and likely to be diseased, were burned wherever they were found.

These back-to-back storms terrified Floridians and shook their confidence. Troubling questions arose. Was the price in human lives and money to make Florida hospitable worthwhile? Had it been a mistake to try to tame the Everglades in the first place?

Another Chance

By now the drainage district was bankrupt and many of its backers demoralized. Death and disaster seemed to be the main legacies of Florida's big gamble to tame nature. Making matters worse, the state government lacked funds, or the will to raise them, to continue paying for drainage projects adequately.

The stalwarts of draining were not discouraged, however. They just lacked resources. "Unable to collect drainage taxes, borrow more money, or meet bond payments, the state turned to federal aid, specifically aid from the Army Corps of Engineers, the only federal agency equipped to undertake such a grandiose task as draining the Everglades," writes Clay J. Landry, research associate for the Property and Environment Research Center, a free-market environmental think tank based in Bozeman, Montana.

In 1929 that federal help arrived. Responsibility for flood control, drainage, and water supply was now in the hands of the U.S. Army Corps of Engineers, whose origin dates back to June 1775, when a chief engineer and two assistants served George Washington.

The Florida legislature acted too, creating the Okeechobee Flood Control District that assumed many of the duties of the Everglades Drainage District.

America's engineers spent the next fifteen years building floodway channels, control gates, and major levees along the shores of Lake Okeechobee. They also erected a much stronger eighty-five-mile-long rampart along three-quarters of the Lake Okeechobee shoreline to keep it from spilling over again. The fortification also came with a new name: "Herbert Hoover Dike," selected to pay homage to the U.S. president, also an engineer, whose political heft helped obtain federal funding for the project.

Floating dredges for the Corps and others soon were scooping up muck all over South Florida. The work was hard. Hired hands endured hunger, thirst, heat, humidity, wild animals, thunderstorms, lonesomeness, and lots of bugs.

"People nowadays can't realize what insects used to be like here in this end of Florida," recalled Lawrence E. Will of his days as an Everglades dredge man. "[W]e'd have sand flies along the ocean beach, horse flies in the spring, and all summer long we had deer flies—and of course, gnats in your eyes, nose and ears . . . there might be smothering swarms of chizzywinks [nonbiting blind mosquitoes] and those flying ants with red hot feet which came in droves on muggy afternoons, but the most numerous, universal and pestiferous of all the insects were the . . . mosquitoes."

Not all dredging was for land reclamation. The federal government, for instance, widened and channeled Florida waterways for navigational purposes in Southwest Florida. Better navigation was also the reason for dredging the Intracoastal Waterway, an inland network of canals, inlets, bays, and rivers that runs along most of the length of Florida's Atlantic coast, and a big stretch of the Gulf coast.

The main goal of dredging at Peace River was to facilitate the conveyance of mined phosphates used in making agricultural fertilizer from the western part of Central Florida known as Bone Valley.

Dredging also produced new areas of dry land. Again and again, scooped-up muck was used as landfill and heaped upon wetlands to make more agricultural fields and homesites. Marshes, bogs, coastlines, estuaries—all were filled with muck, clawed up from a river or bay bottom by huge machines to make islands or reconfigure coastlines and lakefronts.

Into the Wild in Cars

In 1928 an almost epic thirteen-year-long dredge-and-fill road project was finished and ready for traffic—the Tamiami Trail. Workers had dug up muck to provide the roadbed, leaving scooped-out canals along the two-lane ribbon of asphalt that cut east and west through the southern region of the Everglades and the Big Cypress Swamp, connecting Miami on the Atlantic and Naples on the Gulf. In the coming years, the road would be extended to Tampa.

Rightfully hailed as a boon to transportation, the roadway, however, also functioned as a dam that blocked much of the Everglades' flow of fresh water moving south. In the coming years, it became a highway of death for Florida wildlife. Among the tons of splattered roadkill that accumulated over the decades on the highway were corpses of the endangered Florida panther.

As cars and trucks ventured headlong into the Everglades on the new highway, big construction plans were being drawn up for an even bigger transportation project in North Central Florida. This one had international implications.

Florida's Big Ditch

On the morning of September 19, 1935, President Franklin D. Roosevelt sat in his home in Hyde Park, New York, poised to blow something up.

At the scheduled moment, the president pressed a button linked by telegraph wire to set off a dynamite blast more than eleven hundred miles away and nine miles south of Ocala, Florida. Applause and cheers erupted from a crowd of fifteen thousand people who had gathered for the spectacle. Schools had been closed so children could attend. A band played, adding a dose of pomp and circumstance. Everyone present that morning believed they were witnessing a wondrous moment in history.

Such was the spectacular start of a Depression-era federal jobs program to construct a commercial thirty-foot-deep across-state waterway for ships that would run east and west through the heart of Florida. Building such a passage was an old idea dating back centuries to the days of the early Spanish. Now it looked as if it were going to happen.

If all went according to plan, the canal would enable ships to convey their cargo from the Gulf of Mexico near Yankeetown through the middle of Florida all the way to Jacksonville. Shipping experts claimed the trans-Florida route would be safer and cut expenses by shaving three days off the alternative route, which required ships to travel around the Florida Keys. Many U.S. military leaders were confident such a canal would keep American vessels secure from enemy naval attacks. Unemployed workers across the Depression-struck nation also welcomed the coming of the construction project and its promise of jobs.

From the start, however, there was opposition to building the "Big Ditch." Even before the first chunk of earth was dug up, critics voiced concern that construction would damage the Ocklawaha River and its surrounding environment. Resistance also came from railroad companies opposed to competition from shipping interests. On February 10, 1933, their spokesmen argued at a special hearing before the Corps of Engineers in Jacksonville that building the canal was too expensive, and that digging deep into Florida limestone would damage the underground water supply.

The Corps had qualms, too. It was concerned about both the cost and the feasibility of building the canal. Influential politicians and businessmen, however, wanted the job done. Thus, any objections went nowhere, and construction began.

Controversy, however, continued over the coming years, as Floridians watched huge earth-moving equipment and draglines tear up a pristine, beautiful swath of Florida. One notorious machine, known as the Crusher, mowed down huge cypress trees, sometimes as many as six at a time.

In addition to escalating protests from environmental organizations, such as the Florida Defenders of the Environment—a group of local activists who banded together to protect the Ocklawaha River from the perils of the Cross-Florida Barge Canal—the project ran into periodic funding problems. At one point in its uncertain career, the canal was repurposed as a much shallower barge, rather than a ship, canal. By the 1960s, most cargo was being shipped by trucks, not trains or ships. But this mattered little, because on January 19, 1971, President Richard Nixon halted construction of the canal forever, stating publicly, "A natural treasure is involved in the case of the Barge Canal—the Oklawaha [sic] River—a uniquely beautiful, semitropical stream, one of a very few of its kind in the United States, which would be destroyed by construction of the Canal."

Turning to the Feds Again

During the decades Floridians had spent fighting over this ill-fated project, the building of numerous other canals, drainage ditches, and water diversions, plus land-clearing and swamp-filling were altering the state north to south. Dragline operations were steaming away in Southwest Florida, the Panhandle, the Keys, Central Florida, Jacksonville, St. Augustine, and, of course, South Florida. By the 1940s, the so-called land reclamation movement had greatly reduced the size of the Everglades, leaving a wake of destruction that killed wildlife, destroyed native vegetation, and wrecked the natural rhythms of water flow.

Lack of rainfall often made conditions worse. Dried muck, the residue of thousands of years of decomposing saw grass, often caught fire. Black smoke darkened the skies, as cattlemen, citrus growers, and homeowners prayed for relief.

In the summer of 1947, help from the heavens did arrive, but it was too much. First to hit was a hurricane, followed by two tropical storms. Altogether these tempests drenched Central Florida with 100 inches of rain, causing the Caloosahatchee and Kissimmee Rivers, the New River in Ft. Lauderdale, and many other waterways to overflow their banks. The Herbert Hoover Dike held up, but flooding in the Lake Okeechobee region was widespread. Many small dikes that farmers had built collapsed during the intense downpours and flooded tomato and bean crops, ruining them.

"When the sun finally broke through again in November, it shone down on an inland sea—5 million acres of water stretching from Lake Okeechobee across the Everglades and the Big Cypress Swamp to Broward and Dade counties," according to an account in the *Sun-Sentinel.* "Ninety percent of eastern Florida, from Orlando to the Keys, was under water."

Floridians again turned to the nation's capital for help. That assistance came in 1948, when Congress enacted the Flood Control Act and established the Central and Southern Florida (C&SF) Project. Soon, the Army Corps of Engineers began work on new flood control projects from Orlando to the Everglades.

The engineers had three big goals. The first was to build a one-hundred-mile-long level buffer around the east perimeter of the Everglades to contain overflow from Lake Okeechobee and prevent it from ever again flooding out nearby towns and communities. In the coming years, the Corps would also create twenty-six hundred miles of levees and canals, along with dozens of pumping stations and hundreds of water management structures to control flooding.

The second goal called for a big chunk of the northern Everglades—about one-fourth of the entire system—to be set aside for draining and farming. On the south side of Lake Okeechobee, another area was—and still is—designated as the Everglades Agricultural Area (EAA).

The final part of the project focused on protecting the remainder of the Everglades, a region that stretched between the EAA and the 1.5-million-acre Everglades National Park in South Florida. The park was established by Congress in 1947 and was the first ever whose purpose was to preserve animals and plants. Three water conservation areas—huge wetland impoundments—were created to manage the southbound water.

All this work required divided responsibilities. The Corps did the construction work, while the C&SF Flood Control District helped pay for the project and carried out day-to-day operations and maintenance.

On September 9, 1960, much of Florida's new massive outdoor plumbing was severely tested when another hurricane roared ashore. This one had a name, Donna, and it packed winds exceeding 150 miles per hour. The monster storm rode across the Florida Keys and then plowed north through the middle of the state before veering east into the Atlantic. Along the way, Donna inflicted massive flooding in the Tampa Bay area.

The Florida legislature responded to Tampa's distress by authorizing creation of a vast system of levees, canals, and locks to reduce any future flooding in the coming years.

Dredging Frenzy

As governments large and small spent years trying to control flooding and regulate water flows, Florida builders were caught up in a construction frenzy inspired by dredge-and-fill techniques pioneered by Carl Fisher. Across the state, they dredged canals and used the muck to fill in wetlands and create waterfront properties for new homes and commercial buildings. In the process, the developers also deformed topography, hydrology, and natural habitats across the peninsula.

Marshes south of Tampa, along Salt Creek in St. Petersburg, and at the mouth of the Hillsborough River were smothered. Meanwhile, bayside shorelines grew in size at barrier islands in Pinellas, Manatee, and Sarasota Counties. New waterfront properties also appeared in St. Petersburg's Boca Ciega Bay after developers scooped up five thousand acres from the bottom of the bay.

In South Florida many builders were following the example of developer and former West Virginia coal miner Charles Green Rodes, who in the 1920s tore out tons of river bottom from Ft. Lauderdale's New River and used them to create fingers of land placed at a right angle to the river. Every newly built house had its own backyard canal where homeowners could launch boats—just as the wealthy homeowners did up the coast in Palm Beach.

For decades, a "finger frenzy" swept through Florida. "Rodes's finger-island idea proved so profitable, it was copied throughout the state, most notably in Cape Coral, near Fort Myers, where dredge-and-fill work wiped out 18,000 acres of wetland in the 1960s and created some 400 miles of fingers canals along the Caloosahatchee River," write journalists and authors Craig Pittman and Matthew Waite in *Paving Paradise: Florida's Vanishing Wetlands and the Failure of No Net Loss.*

Corps Controversies

The federal government played a major role in this world of dredge and fill by providing the legal framework, funding, and the physical labor to wipe out watery areas. "By the 1960's, most political, financial, and institutional incentives to drain or destroy wetlands were in place," write Thomas E. Dahl, of the U.S. Fish and Wildlife Service, and Gregory J. Allord with the U.S. Geological Survey. "The Federal Government encouraged land drainage and wetland destruction through a variety of legislative and policy instruments. For example, the Watershed Protection and Flood Prevention Act (1954) directly and indirectly increased the drainage of wetlands near flood-control projects."

These two scientists say the federal government abetted wetland destruction through the Department of Agriculture's various "conservation" public-works plans responsible for as many as 550,000 wetlands acres lost annually from the mid-1950s to the mid-1970s.

Among the many federally backed drainage projects, one stands out for the colossal ecological disaster it inflicted upon Florida's freshwater systems. This boondoggle began in the 1960s, when the Corps, working with the South Florida Water Management District, commenced a massive ditch-and-drain project to straighten the slow-moving, meandering Kissimmee River in Central Florida, which traveled 103 miles from Lake Kissimmee south to Lake Okeechobee and made a floodplain that in some places was three miles wide.

Ignoring early warnings from some scientists, the Corps spent a decade reshaping the river into a fifty-six-mile-long canal, three hundred feet wide and thirty feet deep. In the process, the Kissimmee lost not only its integrity but also its Indian name. Now it was called something only engineers could have fancied: the C-38 Canal.

In 1971 the project was completed. In a sense it was a success. Hadn't the engineers accomplished the goal of reducing flooding and opening up more dry land, especially for ranchers?

But C-38 infuriated many people. Complaints poured in from around the nation. By now, a growing number of Americans understood that wetlands were important components of Florida's hydrology and ecology.

Flashes of enlightenment also flickered even in chambers of government. In fact, by this time, Congress was replacing laws it had crafted to destroy wetlands with new ones that authorized just the opposite, such as the Emergency Wetland Resources Act of 1986.

By 1991, even the Corps itself agreed the disfigurement of the Kissimmee was a colossal mistake. Then with new authorization from Congress, the army engineers got busy repairing the first-magnitude damage they'd inflicted on Florida.

Their new goal? To restore wetlands and de-canalize the Kissimmee. It was now understood that the original twisting, serpentine shape of the river was a good thing. It had helped to filter pollutants and slow down the southward flow of water to Lake Okeechobee. It provided wildlife habitat. And it was pretty.

Today, the Corps has nearly finished that historic restoration. It's been an enormous undertaking—a tedious, immense, $1-billion restoration of the river's historic shape and flow.

During those decades, the Corps was also involved in many other projects. In fact, in modern times it has always been inseparable from the fate of Florida's wetlands. Federal law gives the Army Corps alone the sole authority to grant permits to developers that only the EPA can override—something that seldom happens.

The Right to Destroy Wetlands

It's also rare the Corps denies a wetland permit in Florida. "Federal regulations stipulate that authorization shall be granted unless a project is determined to be contrary to the public interest, and the Corps weighs the rights of private property ownership in consideration of the Fifth

Amendment to the Constitution," explains Nakeir Nobles, a spokesperson for the Corps' Jacksonville office.

But how does this concern for property rights square with the Clean Water Act, the nation's number-one water law? After all, the declared goal of the act "is to restore and maintain the chemical, physical, and biological integrity of the Nation's waters."

And what does "public interest" mean? To an environmental activist this term might suggest the correctness in keeping a Florida wetland healthy and functioning to cleanse water of contaminants.

Developers, on the other hand, are likely to say an outlet mall they want to build where a wetland now stands will create new jobs and boost the county's land taxes. What could be more in the public interest than that?

The issue gets pricklier. For years, those in America's property-rights movement have asserted that when the Corps doesn't allow wetland owners to convert their private wetland properties to a more profitable use, that property loses market value. In effect, they say, the government has taken some of their land's value.

Here's where the fifth amendment enters the story. Its "takings clause" reads: "[N]or shall private property be taken for public use, without just compensation."

Therefore, if there's no compensation for these market losses, say property rights defenders, wetlands owners should receive approvals of their permits.

However, their opponents often reply that wetlands also provide public benefits such as water filtration, wildlife habitat, and buffering against flooding and erosion. Wetland owners, they add, also receive these same benefits, which should offset any losses associated with takings. Besides, why should wetland owners be compensated simply for not inflicting injury on nature?

Not so fast, respond would-be wetlands developers. The public enjoys these benefits at a cost to us: that's because these wetlands are desirable places and are kept that way at our expense. It's not fair.

"In Congress, the 'property rights issue' has played out with particular force in the area of wetlands regulation . . . in part because of polarized views on how private property rights should be accommodated," writes Robert Meltz, an attorney with the U.S. Congressional Research Service. "The result is that the wetlands permitting program . . . gives the administering agencies (the Army Corps of Engineers and Environmental Protection Agency) little in the way of clear guidance."

There is also disagreement over what a wetland is. The EPA defines one this way: "Wetlands are areas where water covers the soil, or is present either at or near the surface of the soil all year or for varying periods of time during the year, including during the growing season."

A century ago, many Floridians would have thought almost any standing pool of water was a waste of space. Modern biologists, however, are more likely to find compelling ecological reasons why keeping a wetland unmolested makes sense. First of all, they say, wetlands are not watery wastelands. Instead, they function as natural water pollution filters, wildlife habitats, hurricane barriers, and freshwater recharge areas. An acre of wetland may soak and store up to 1.5 million gallons of water—a useful thing to have during a flood.

No Loss

For decades the Corps has approved most of the permits that have arrived at its Jacksonville office. However, today it also recognizes the environmental importance of wetlands and has adopted a plan for dealing with them. The policy, initially embraced by the George H. W. Bush administration in 1989, is based on the idea of "no net loss" of wetlands. This approach is supposed to satisfy the desires of the permit applicants and the needs of nature. Advocates call it a win-win approach. Skeptics don't.

It works like this: Suppose a developer wants to fill in one hundred acres of wetland. To get a permit, he or she must agree to offset the damage by creating new or restoring existing wetlands in the same water basin.

Or the developer can use the services of a third party—a mitigation banker—who has allegedly converted an old pasture or pulped forest into a wetland. If this land has specific environmental values that match those of the land being destroyed by the developer—and if federal and state regulators approve the deal—the developer can then buy a certain number of credits from the mitigation banker, who will swap part of his or her artificial "wetland" land for the real wetland being wiped out.

Under Corps policy, it has been possible in recent years for Florida developers to get preservation credit for wetlands they don't destroy.

Imagine, for example, that you're a developer with one hundred acres of wetlands in Collier County. You could file a permit to develop seventy acres, and in return you agree to set aside the remaining thirty acres for "conservation" purposes. With an approving nod from the Corps, you could now claim that you've saved thirty acres from destruction—or euphemistically that you "preserved" wetlands.

Critics of such a policy say this is akin to telling a little boy that you are pleased with him because his roughhousing broke only one of his mother's cherished twelve dinner dishes. After all, he preserved eleven!

The Florida Department of Transportation deployed this kind of logic to justify constructing the forty-two-mile Suncoast Parkway that destroyed at least two hundred acres of wetlands in Hernando, Pasco, and Hillsborough Counties. To make up for the damage, DOT bought ten thousand acres of undeveloped land and promised the Florida public that roughly a third of the property, which was wetlands, wouldn't be bulldozed.

It almost sounded like Florida's environment was getting a bonus. But it wasn't. The two hundred acres of wetlands are gone forever. Only a promise not to destroy about three thousand acres is in place. And that promise may be waning. In late December 2017, Pasco County government was plunging ahead with plans to build a road through the preserve.

A similar argument was at the heart of a legal case whose final chapter may force government agencies to go easier on wetland developers in the future. In 1994 the late Coy Koontz requested a wetland development permit from the St. Johns River WMD to develop 3.7 acres on about a 15-acre parcel in Orange County, Florida. He also offered to set aside the remainder for conservation purposes. The district wasn't satisfied and told Koontz that to obtain the permit, he'd have to pay for wetlands protection on district land not far away. Koontz refused and sued, claiming the cost was excessive and an unlawful taking.

The case wended its way through the judicial system. Finally, in January 2014, the U.S. Supreme Court with a 5–4 decision ruled in Koontz's favor.

Does It Work?

Advocates of "no net loss" have touted it for years as a good idea. But how effective is it? The investigative reporters Pittman and Waite provided an answer in a May 23, 2005, *St. Petersburg Times* news article: "[S]ince the policy took effect in 1990, at least 84,000 acres of Florida wetlands have disappeared . . . The corps approves more permits to destroy wetlands in Florida than any other state, and allows a higher percentage of destruction in Florida than nationally. Between 1999 and 2003, it approved more than 12,000 wetland permits and rejected one."

It's not clear that the no-net-loss policy results in an even trade-off. In *Paving Paradise*, Pittman and Waite reported that all too often Florida's artificial wetlands are lifeless holes in the ground that sometimes fill with water and often don't. Many are just rock pits, say the journalists, or ponds stuck in the middle of highly fertilized golf courses. Too many are located in places where they aren't needed. Some are dry much of the year. Says Pittman, "Florida lets the bankers count dry land as if it were wet, thus totally flipping the idea of wetland mitigation and no net loss upside down."

The key takeaway from the reporters' investigations is that few of the artificial wetlands fully restore the environmental values that were lost.

There are problems too even when dry land isn't part of the deal. Wildlife, of course, is damaged when wetlands are wiped out. People in the area losing wetlands also forfeit the benefit of those wetlands. "If the wetlands move, their ecosystem services go with them," write wetlands scholars J. B. Ruhl and James Salzman.

Whether newly created wetlands at the receiving site will be as productive in the future is another uncertainty. And will they function the same way as the ones being destroyed? Such considerations today are supposed to be taken into account with modern wetland swaps.

No End to Dredging

Though rules are tougher today, Florida wetlands continue being wiped out and altered by dredging and filling operations to build homes, lay down pipelines and cables, and improve navigation.

There are also channel-deepening projects, which are often controversial. For instance, by early 2017, it looked like it was all-systems-go for launching a $684-million project in Jacksonville to deepen a thirteen-mile ship channel segment of the St. Johns River from forty to forty-seven feet to accommodate bigger container ships now passing through the Panama Canal. The St. Johns Riverkeeper opposed the dredging, fearing a deeper channel will increase salinity and pollution in the river, threaten wildlife, and intensify sea level rise. Similar concerns have been raised in Miami and other coastal cities over channel-dredging projects.

In recent years, there's been turmoil also in Key West, where residents are divided over whether to allow dredging of a ship channel to allow ever-larger cruise ships to make port there.

At the dawn of the twentieth century, controversy over dredge and fill was typically over what methods worked best. Many Floridians favored the practice, no matter how it was done. After all, it provided dry land, made transportation safer, provided flood protection, and reduced a huge amount of wet soil that mosquitoes needed to reproduce and thus made life more endurable and desirable for humans in Florida.

However, Floridians today understand—or should understand—that for all the good done by dredge-and-fill operations of the past, they also destroyed much of the natural world upon which our water supplies depend.

And wetlands still get in the way of development.

In February 2018 prospects for speeding up development looked brighter to Florida developers. That's because Governor Rick Scott and the Trump administration negotiated a plan to turn over the Corps' wetlands-permitting responsibility to the Florida DEP.

Florida lawmakers, meanwhile, were doing their best to make this change happen. Backers of the idea said the DEP was up to the job, and it would reduce redundancy and confusion in the permitting process.

Environmental communities were divided over the issue. None of them had ever been happy with the Corps' handling of wetlands in the past. Many activists, however, also questioned whether a DEP that was still hard hit with massive layoffs and a tight budget would be able to carry out the burden of wetlands permitting. Some also worried the department was still not free of political influence from powerful politicians. How could DEP staffers, they wondered, handle the extra workload and do it correctly while being under pressure to speed up wetland protection decision making? Another of their objections was that no new funding was coming from Tallahassee to help DEP take on the new tasks.

The Florida Chamber of Commerce's position, however, was clear and predictable. According to Christopher Emmanuel, the pro-development organization's director of infrastructure and governance policy, the chamber backed the proposed legislation, which "follows similar successful state efforts to assume federal permitting responsibilities for wastewater discharges and air emissions."

In early March 2018 the Florida legislature passed a controversial bill setting up the legal framework at the state level for a possible transfer of coveted wetland permitting authority. At the time of this writing, the EPA and the Corps were considering Florida's request.

As Floridians continue to wrangle over both this new law and the legacies of past water policies, another double-edged force is tearing up the state. It's the Florida growth machine. And fresh water is its lifeblood.

6

The Florida Growth Machine

FLORIDA IS ONE OF THE NATION'S LEADERS!

In water guzzling, that is.

"As in 2005, water withdrawals in four states—California, Texas, Idaho, and Florida—accounted for more than one-quarter of all fresh and saline water withdrawn in the United States in 2010," according to a U.S. Geological Survey (USGS) report. Florida is also the biggest consumer of groundwater of all the states east of the Mississippi River.

Of course, one obvious explanation for Florida's big thirst is that it is the third most populated state. It also has the fourth-largest economy in the country. Only those of California, Texas, and New York are bigger. Tourism, aerospace and aviation, life sciences, financial services, the military, international trade, mining, agriculture, construction, and, by some economic reckonings, the international illegal drug trade are the big economic drivers in Florida. Fueling every one of these enterprises is relentless population growth. All require fresh water.

"The Sunshine State has never been a place for slow-paced growth or the quiet process of augmentation," writes author David Nolan in his book *Fifty Feet in Paradise: The Booming of Florida.* "It has, rather, followed cycle after cycle of hopes, hypes, dizziness, collapse, forgetfulness, and rebirth."

One of the biggest of those cycles—akin to a growth spurt for a puberty-stricken youth—began at the end of the Second World War.

The United States, triumphant and vital, was now a superpower armed with nuclear weapons and a jaw-dropping economic clout to match. At home, legions of returning U.S. G.I.s, weary of war, wanted to marry and start families. And there was one very special and inviting sunny place that beckoned.

For one thing, many returning veterans remembered being stationed in Florida for training purposes. They liked what they saw and returned after the war, towing a spouse and kids. Making the adventure even more inviting were fistfuls of generous benefits from the federal government to help them begin a new life. The still sparsely populated land of sunshine seemed just the place!

"Now the state was set for a new boom, which began as soon as the war ended," Nolan continues. "The flood of tourists was so great that many had to sleep in their cars, while the demand for homesites was sufficient to quadruple the price of building lots. VA-FHA-guaranteed mortgages fueled the flames. Florida was off again!"

Older Americans were moving to Florida, too. In previous decades, retirees had usually finished their days in their hometowns, getting by on reduced funds and the help of nearby relatives and friends.

By the late 1940s, however, older Americans had options. They now had Social Security checks that allowed them to live independently. So why not live in sunny, warm Florida, at least part of the year? The cost of living down there was cheap. There was no state income tax. Besides, many of the retirees' grown children were also relocating there.

To house everyone, developers cleared Florida scrub and used the same blueprint over and over again to build affordable, cookie-cutter homes in vast, monotonous subdivisions. Land taxes were cheap, too, in part because in 1934 Florida voters amended the state constitution to exempt property taxes on the first five thousand dollars of the value of a permanent residence.

Enticed by these incentives, Americans by the millions migrated south. The influx included a vast array of humanity, including an assortment of carpetbaggers, schemers, opportunists, criminals, and wheeler-dealers hoping to exploit and cash in. The sweet smell of easy money from the land boom days of the 1920s had returned to the subtropical air.

As author Hodding Carter notes, "Since Florida became a state, all it's been is 'What's in it for me?' I know this is our modern-day mantra, but in Florida it's gone beyond a mantra—it's the whole religion. And the place is peopled with zealots who are blinded by their bigotry and distrust of everybody else. It's really getting me, pulling me down, down in the Everglades muck."

Seeking Paradise

Like so many who had moved to Florida in the past, many of Florida's eager new transplants sought a modern Utopia. After all, for more than a century, Florida promoters had portrayed the state as a "Garden of Eden," "Tropical Paradise," "Promised Land," or "Fountain of Youth."

Even today, many Americans still see Florida as a place to escape to. Unemployed? Divorced? Depressed? Don't worry, say Florida's savviest beguilers. Simply relocate to the land of dreams filled with tons of beaches, barbeques, and parties, where fruit grows on trees in winter. You, too, can have fun sipping a margarita, right alongside Jimmy Buffett.

Is there something baked into the human psyche that makes it easier for promoters to pitch Florida as a wondrous garden? Maybe so, says E. O. Wilson, the preeminent Harvard biologist and naturalist. In his 2014 book, *The Meaning of Human Existence*, Wilson cites a study of volunteers from different countries who were asked to look over photographs of various habitats, ranging from forests to deserts. Then they were asked to pick the ideal place to live. "The preferred choice," Wilson recounts, "had three factors: the ideal vantage point is on a rise looking down, a vista of parkland comprising grassland sprinkled with trees and copses, and proximity to a body of water, whether, stream, pond, lake, or ocean."

This archetype, says Wilson, "happens to be close to the actual savannas of Africa where our prehuman and early ancestors evolved over millions of years."

Does that mean we humans have a gene-coded preference for a landscape resembling where our ancestors came into being? It's not a far-fetched idea, argues Wilson. "All mobile animal species, from the tiniest insect to elephant and lions, instinctively choose the habitats to which all the rest of their biology is best adapted."

Uber-flat Florida, of course, doesn't have many rises. But imagine a couple from Boise, relaxing with drinks on the balcony on the third floor of their Ft. Myers condo, overlooking a sunny golf course with rolling green turf, palm trees, and blue ponds, and you sense what might have spurred this couple to pack their household goods one cold, wintry night and head off to Florida.

Come on Down

Promoters have also always tried to advertise Florida as the ultimately cool, or hip, place to visit. Often, celebrities have been recruited to carry out this dubious task. In 1875, for instance, the Atlantic Coast Line Rail Road hired poet and former Confederate soldier Sidney Lanier to write a booklet extolling the wonders of Florida to encourage Americans to ride its trains down yonder.

During the Florida land boom of the 1920s, the populist, statesman, and defender of the Holy Bible in the Scopes "Monkey Trial," William Jennings Bryan, considered one of the nation's most brilliant orators, applied his gift of gab to flogging real estate and tourism in South Florida.

In the 1950s, television arrived and among its early broadcasts were black and white images of a ukulele-strumming crooner and promoter named Arthur Godfrey performing in a Miami Beach hotel. "Come on down!" to Florida, the folksy "Old Red Head" implored his viewers.

A few decades later, Florida's marketing image was shaped anew by Anita Bryant, a former Miss Oklahoma beauty pageant winner (and later antigay activist), who sang her way through citrus groves, touting Florida orange juice, until her public homophobia soured her standing with the citrus industry.

In 1994, however, the state Citrus Commission found someone new: conservative radio commentator Rush Limbaugh—known widely for bashing feminists, liberals, and environmentalists—to peddle Florida orange juice on his radio show for $1 million. The deal, though, mostly succeeded in generating outrage from liberals.

The most recent celebrity to make a pitch is Miami-based hip-hop rapper Armando Christian Perez, better known as Pitbull. In 2016 controversy flared up again when the public learned that a for-profit marketing company, Visit Florida, on behalf of the Florida state government, secretly paid Pitbull $1 million in taxpayer money for his promotional video "Sexy Beaches."

There it was! On social media, images of pouting, promiscuous young women in thongs, drinking, shop-lifting, and kissing one another, promised, "We will do all the wrong things/Wrong things you like." The video was supposed to reel in millennials by the truckloads to party-loving Florida.

No lucrative state contracts have yet emerged, however, with celebrities who urge the conservation and restoration of Florida's natural resources now under constant assault from growing crowds of visitors enticed to the state's commercial interests.

The Floridian Who Really Made Florida Cool

It's probable most of the savanna-seeking pilgrims in the past half century never would have headed south on Dixie Highway, if it hadn't been for the hard work of a Florida Panhandle doctor who tried to make Florida cool—in a literal sense. His likeness appears among the life-size marble statues of notable Americans arrayed in the National Statuary Hall in the U.S. Capitol, representing people of honor from each state.

As every Floridian should know, one of the two people chosen to represent Florida was John Gorrie. In 1833 this South Carolinian polymath settled in Apalachicola, then a busy cotton port on Florida's Gulf coast, where he served as town mayor and assistant postmaster. Gorrie also founded a church and became a practicing physician, inventor, and scientist.

In addition, the busy doctor added his voice to the chorus of Floridians calling for intense water drainage; however, he wasn't interested in reclaiming any wetlands. Instead, Gorrie's focus was on public health. Drying up swamps, he believed, would help to wipe out mosquito habitat and thereby reduce the number of malarial cases he had to contend with.

Concern for his patients, in fact, inspired him to create a contraption that could artificially cool a room sufficiently enough to comfort his feverish malarial patients. Gorrie's design helped pave the way for modern air-conditioning, which by the 1960s was appearing with increasing frequency in Florida stores, schools, offices, homes, and cars, making the sultry, humid state bearable. You could even watch a movie at an outdoor air-conditioned drive-in movie theater in Miami.

Baby, You Can Drive My Car

The drive-in wasn't the only emblem of the emerging car culture. Suburbs were popping up everywhere, built for the convenience of motorists. So, too, were a growing number of motels, fast-food restaurants, and shopping centers. To accommodate Florida's ever-multiplying number of new roadsters, construction crews were busy paving a vast web of roads, highways, and superhighways across the state.

During the 1950s, Congress broadened its own road-building presence when it authorized construction of the Eisenhower Interstate Highway system. Its foremost purpose was to provide a major road system that the U.S. military could use in emergencies to travel from one side of America to the next without meeting a red light, presumably to blast away invading communist armies.

In the coming decades, federal superhighways also made it easier for families and tourists—and criminals, too, say sociologists—to travel to Florida and zip around within the state once they'd arrived. I-95 funneled in migrants and tourists alike from the northeastern and mid-Atlantic states along Florida's east coast. Meanwhile, I-75 allowed retirees and others from the Midwest to travel down to the center of the state and the west coast. Later, when the east-west-oriented I-10 segments snapped into place, even more people in automobiles and trucks traveled the state quicker than ever before.

Beginning in the late 1950s, other newcomers began showing up. They were Cubans, escaping revolution and the despotic rule of Fidel Castro. Hundreds of thousands more would follow and settle in South Florida.

Haitians, Jamaicans, Bahamians, and people from Central and South America also arrived by the tens of thousands. For the past half century, in fact, Florida has been attracting people from around the globe. They made the trips in homemade rafts, cars, trucks, and jet planes—sometimes a thousand arriving every day. Today, the Miami-Dade School system enrolls students who were born in more than 160 countries. All of them needed space, food, shelter—and water.

More people were yet to come.

The Mysterious Stranger

On November 22, 1963, the same day an assassin murdered President John Kennedy in Dallas, Texas, a small plane flew over Central Florida. On board was a rich, successful American man who burned with his own special kind of empire fever. Unlike other business tycoons who'd come to Florida, this one had artistic flair, and he wasn't interested in building on the Atlantic coast. Instead, he wanted a site somewhere down below where the plane's shadow was racing across the sparsely populated citrus groves, scrubby ranch lands, and buggy swamps of Central Florida. At last, he cried, "This is it!"

Not long afterward, the mysterious air traveler launched a secret mission, code-named Project X. Among his agents were men who had conducted U.S. military intelligence operations during the Second World War. They quietly fanned out across a huge swath of Central Florida territory southwest of Orlando and bought up properties big and small as fast as they could. The average purchase amount was $180 an acre. Altogether, the men bought a total of 27,443 acres, or 43 square miles, of Central Florida. It was a chunk of real estate about twice as big as Manhattan.

The secrecy paid off. Had the sellers known who was behind Project X, they likely would have hiked their asking prices. The secretive land grab hadn't gone unnoticed, however. Orlando was abuzz with rumors.

Finally, the truth came out. On October 24, 1965, this headline appeared in the Sunday edition of the *Orlando Sentinel*: "We Say: 'Mystery' Industry Is Disney."

Florida governor Haydon Burns followed up with a press conference, along with the furtive man himself and his brother Roy. He was, of course, Walt Disney, who took advantage of the public event to reveal that he wanted to build both a huge theme park, bigger than Disneyland in California, and a futuristic city just outside Orlando.

The governor at his side publicly effused that Disney was "the man of the decade who will bring a new world of entertainment, pleasure and economic development to the State of Florida."

Land prices dutifully exploded. Orlando's business community was thrilled. At last, their sleepy, scrubby, lake-dotted Orlando was going to be transformed into an important, international city!

Disney's business now emerged from hiding and into the daylight, calling itself the Reedy Creek Ranch. After opening an office in downtown Orlando, the company went to circuit court and obtained permission to become its own "drainage district." This meant Reedy Creek could create and control its own water system, entitled to sell tax-exempt bonds to pay for its expenses.

That was just the beginning. Disney's legal team next obtained special approval from a fawning state legislature to legally transform itself again, this time into an "improvement district." Only later did state lawmakers take time to read the details tucked away in the legislation, written by Disney's own attorneys. Too late they discovered they'd given the Disney organization the right to add more lands into their district, avoid paying property taxes, and get tax refunds. Disney also had the authority to create two "municipalities," Lake Buena Vista and Bay Lake, both built on the banks of reservoirs, created by damming the natural flow of water.

These Disney "cities" are unlike any others in Florida. They are run by a corporation, not a citizenry. "Never before or since has such outlandish dominion been given to a private corporation," complains novelist, columnist, and investigative reporter Carl Hiassen in his nonfiction work *Team Rodent: How Disney Devours the World*.

As an improvement district, the Disney Corporation enjoys almost all the powers of authentic Florida cities. In addition to owning and operating its own water and sewer systems, it can build and operate its own electric power plants, levy taxes and issue bonds, impose eminent domain to seize property belonging to others within the district, and establish its own building codes enforced by its own inspectors. Disney also maintains its own fire department and a security force that at times acts as if it were a genuine law-enforcement agency.

A five-member board of directors governs the Reedy Creek Improvement District (RCID). Its provisions for holding elected office are unique. "According to RCID's special act, an individual must own land within the district in order to serve on the board . . . The district's special act provides that at elections of supervisors, each landowner is entitled to one vote for each acre of land owned; as the largest landowner, the Walt Disney World Co. is entitled to the most votes," according to an assessment by Florida's Office of Program Policy Analysis and Government Accountability.

Disney "was creating a sort of Vatican with Mouse ears: a city-state within the larger state of Florida, controlled by the company yet enjoying regulatory powers reserved by law for popularly elected governments," writes Richard Foglesong, a professor of politics at Rollins College in Winter Park, Florida.

The Disney model provided a template for other land developers who also want to function as private governments in Florida. Today, Florida has more than six hundred Community Development Districts (CDDs)—special-purpose taxing and development districts created by the state legislature in 1980.

It wasn't until three years after Disney's death from lung cancer in 1969 that construction for Walt Disney World began. Thousands of workers moved earth, filled sinkholes, chopped down trees, and poured concrete and asphalt.

Their impact on Florida's water resources was extensive. "Overall, Disney crews dredged, blasted, constructed, and raised 40 miles of canals, 19 miles of levees, more than a dozen flood-control structures, and 8 million cubic yards of earth," writes University of South Florida historian Gary R. Mormino in *Land of Sunshine, State of Dreams: A Social History of Modern Florida.*

Disney's engineers used their canals to control the water level and keep a reserve for future use. David Koenig, author of *Realityland: True-Life Adventures at Walt Disney World,* explains: "As water rose in one section, gates would automatically float open to release water to the next section, then automatically close when the water level subsided. The 44 miles of winding canals followed the natural curves of the landscape, to blend in."

The state of Florida had given "Disney the power to disrupt Florida's ecosystem with impunity," asserts author T. D. Allman. "If a creek flowing down to the Everglades happened to traverse the theme park, Disney's landscapers now could block its flow, no matter how severe or far-flung the consequences. Damming up these natural waterways allowed Disney to create the artificial lakes that dominate Disney World's landscaping."

In Florida, state and local regulatory agencies have authority only over waters flowing in and out of drainage districts. Thus, Disney was free from "environmental accountability," argues Allman.

However, it is also true that the Disney Corporation has tried to show that it has been going "green" by setting aside some of its lands as preserves and open spaces. It has also adopted many green practices within its theme parks and set up conservation ponds to reduce pollution.

But it is Disney's growing economic impact, not its professed commitment to Florida's environment, that makes it the darling of developers and politicians. According to a 2011 study commissioned by Disney, the company's theme park operations make an $18.2-billion impact on the Florida economy every year. The corporation also claims its economic ripple effect accounts for at least one of every fifty jobs.

So far, there's every reason to believe much of the same will prevail far into the future. For one thing, the Florida Turnpike, which opened January 25, 1957, and I-4, built in the 1960s, today continue allowing hundreds of millions of motorists quick and easy access to the wonderful world of Disney in Orlando.

Midcentury Growth Spurt

Many other big players also contributed to Florida's growth. One of them was billionaire Ed Ball—the potent force within the "Pork Chop Gang," a coalition of Florida good-old-boy segregationists, politicians, and businessmen that dominated state politics for decades. In the 1930s Ball took control of the St. Joe land development company that was slowly changing the face of the Panhandle. Manufactured homes from Tampa-based Jim Walter also popped up everywhere. The General Development Corporation—which once claimed to be the biggest builder in the world—plastered Central Florida, South Florida, and elsewhere with affordable, look-alike suburbs.

Florida's growth rate accelerated in the 1950s with the arrival of the National Aeronautical and Space Administration (NASA) on the sandy coast of Brevard County. Within just a few years the agency launched communications and spy satellites, put a man on the moon, built a space station and pitched it into perpetual orbit, and blasted off space shuttles. NASA's big-league contractors and other support corporations showed up in Central Florida, too. Martin Marietta, for example, built an Orlando plant, employing more than ten thousand workers.

For much of the second half of the twentieth century, other large-scale developments and real estate corporations along with numerous local builders also joined the building spree, carpeting Central and

South Florida, the Jacksonville area, the I-4 corridor, and much of the state's west coast with houses and commercial buildings.

Accompanying all this construction was extensive bulldozing of forests, draining of wetlands, and filling in of swamps. Code regulations were often lax or unenforced. Some counties, in fact, still had no zoning laws.

During much of the second half of the twentieth century and into the early twenty-first, demand for Florida housing rocketed. Regulation in many areas was weak at best. Another plus for developers was that Florida was a "right-to-work" state, meaning it wasn't heavily unionized, so labor was cheap.

And there was plenty of inexpensive land—even if much of it wasn't all that inhabitable. But as Florida developers have long understood, you can alter the landscape with enough draglines and bulldozers.

In 1957 that's what two brothers from Baltimore, Maryland, Leonard and Jack Rosen, had in mind, when they arrived in Southwest Florida's Lee County. There they bought land on a scrubby, swampy, mangrove-rich peninsula known as Redfish Point, located across the Caloosahatchee River near present-day Ft. Myers, and launched a massive housing project called Cape Coral. Over the years, their construction workers drained the swamps by digging four hundred canals—the most of any city on earth—to create "waterfront properties" for incoming hordes of people eager to enjoy some of the "Legendary Lazy Living" promised by Rosen brothers' brochures.

Today, 180,000 people live in a crowded, buggy, dredged-up, flood-prone, hurricane-vulnerable swath of coastal Florida. And another 180,000 paradise-seeking people are expected to join them during the next twenty years.

A Changed World

Florida's growth represents one of the most astonishing relocations of human beings in history. "The decades following 1940 changed Florida more than the previous four centuries, altering boundaries, reconfiguring landscapes, and casting new relationships," writes the historian Mormino. "The march to and across Florida was irresistible and irrepressible, as orange trees became gated communities, small towns were transformed into cities, and big cities into metropolises and boomburbs."

Picturesque Old Florida with its town squares, red-brick courthouses, Spanish-moss-draped live oaks, white-washed Florida cracker houses, and citrus groves on the outskirts of town gave way to soulless strip malls, fast-food restaurants, big-box retailers, and squat, sealed-up houses connected by black asphalt.

At the dawn of the twentieth century, a majority of Floridians were Anglo Saxon/Celtic Protestants who outnumbered African Americans six to one. Florida was also home to Jim Crow racial segregation laws.

But growth was ushering in people with a vast array of backgrounds, values, and beliefs from everywhere on the planet. Florida's days as the least-populated and one of the most culturally backward of Southern states were coming to an end. In fact, by the 1970s, the only thing remotely still southern about much of South Florida was its geography. It was now a medley of accents and a tapestry of races, religions, classes, and ethnicities. Many Floridians with long memories of the state and old Florida accents found themselves a minority, except for those living in remote rural areas, or near the Georgia and Alabama borders. Demographers now spoke of Florida as a megastate with megatrends, as it commanded growing news coverage at the national level.

Growth Is Good?

"In God We Trust" is Florida's official motto. Cynics might offer a companion maxim: "In Growth We Believe." After all, growth has been the operating principle of the state since its inception. Proponents argue that growth creates more demand for goods and services and drives up wages. It also guarantees Florida more congressional seats in Washington. Today, Florida has twenty-nine electoral votes; only California and Texas have more.

However, a growing number of Floridians have raised questions about the assumed blessings of growth. Who pays for the services needed for the newly arrived? Wages go up with growth, but don't prices and taxes also? Doesn't crime increase, too? Is the destruction to the natural world a fair trade-off for more malls and golf courses?

To even express such concerns in some quarters, however, was and still is risky. Just ask Anne Scott, a former Martin County commissioner who says she learned this lesson the hard way. A former attorney and trial judge from Chicago who moved to the city of Stuart, Scott was asked by her neighbors to run for the Martin County Commission. She did and won. But she wasn't reelected in 2016. The reason for her defeat, she believes, is that she didn't embrace the pro-growth agenda set by the local power elite.

"I feel they targeted me on day one, when I became county commissioner in 2012," she says. "Too often, the question around here is reduced to someone saying, 'I'm pro-growth and you are anti-growth.' And I'd always reply, 'What are you talking about? Do you mean population growth? Economic growth? Quality of life growth?' Deciding what is meant by growth is the heart of the whole question."

Scott's troubles haven't ended. Even out of office she's been embroiled in a legal fight rooted tangentially to strife over local development.

The Threat of a Pushback

Scott also isn't alone in questioning the state's Johnny-One-Note paean to growth. As early as the 1970s, in fact, an anti-growth movement had been gaining momentum. More and more Floridians were clearly worried that growth was destroying what made Florida attractive in the first place.

As if on cue, a figure appeared on the political landscape who was prepared, and undeniably competent, to stand up to the Florida growth machine. He was a fifth-generation Floridian named John DeGrove. His admirers still call him Florida's "Father of Growth Management." A St. Augustine native, DeGrove fought in the Second World War, earning both a Silver Star and a Purple Heart. Holding degrees in history, political science, and public administration, he later taught at the University of North Carolina, University of Florida, and Florida Atlantic University, where he was christened Eminent Scholar Emeritus in Growth Management and Development in the College for Design and Social Inquiry.

Through his writings and speeches, DeGrove argued that the state and local governments must adopt policies that implemented rational ways of managing Florida's growth. Without them, he warned, waste and ruin were inevitable.

In 1971, environmentally attuned Florida governor Reubin Askew, whose reputation for personal integrity was known across America, took DeGrove's advice and persuaded the legislature to establish five water management districts. He also convinced lawmakers to make these districts accountable to the governor to keep future lawmakers from meddling in water policy.

DeGrove helped write Florida's 1985 Growth Management Act and State Comprehensive Plan. He perhaps is best remembered for serving as secretary of the Florida Department of Community Affairs (DCA), created by the legislature to manage Florida's runaway growth by making sure proposed large-scale developments were compatible with a region's infrastructure and didn't cause sprawl or worsen transportation problems.

For a while, DeGrove's ideas were hot. They were ballyhooed nationwide as exemplars for the rest of the nation. Year after year, his prestige grew until he was widely acknowledged as the Florida sage of rational and sensible growth management.

Not everyone praised him, however. Many within the development fraternity vilified his efforts, complaining that state regulation tied developers' hands. One critic even publicly compared the DCA to a Soviet Union gulag.

Meanwhile, many elected officials at the local level who received campaign contributions from developers were busily sabotaging their own communities' crafted comprehensive growth-management plans, which they had been required by state law to create.

It was easy. If developers didn't like a restriction on how many houses they could build on a certain tract of land, they had only to ask a county commission to "amend" the land use element of the "comp plan." If a majority of commissioners agreed, as they did thousands of times across the state, developers got what they wanted and made a mockery of DeGrove's philosophy.

St. Petersburg Times columnist Howard Troxler, a longtime critic of how Florida growth management was being subverted, summed up what was going on in a 2010 luncheon speech to the Hillsborough County League of Women Voters: "[It] hasn't done that much good since what happens is: you go through all this planning, and you have technical people, you have experts, you have public hearings, and then the developer gives, you know, a few thousand dollars to the county commission, and they approve it [a land use change]. And that's how the planning process works in Florida."

Many Floridians felt Troxler's pain. Fed up with nonstop growth, urban sprawl, strained water resources, environmental degradation, and what they viewed as political corruption, they joined a statewide grassroots movement called Hometown Democracy. The group's goal was to convince Floridians to pass a constitutional measure then popularly known as Amendment 4. If passed, this change to the state constitution would require voter approval only when local officials wanted to alter existing growth-management plans.

The big question was: could citizens govern themselves? Early polls showed more than 70 percent of Floridians thought they could.

Counter Revolution

The Florida growth machine, however, hated the idea of real democracy and went to war to stop it. As one Amendment 4 opponent put it: "Right now I only have to sway three out of five county commissioners to change the comp plan; I don't want to take on the whole community."

Developers and their allies blitzed the media with negative ads against the proposal. Hometown Democracy would cause unemployment, they cried. The people couldn't be trusted to make wise decisions. Only elected officials could. Letting people vote on the fate of an individual's land was socialism!

The smear job worked. In November 2010, inundated with fear-mongering ads, Florida voters overwhelmingly rejected Amendment 4. Its critics beamed.

The jubilation didn't last long. After all, by now the Great Recession was under way. The housing market had tanked in Florida, across the nation, and in much of the world. At home, yet another Florida housing bubble had burst. Financial institutions, realtors, and construction businesses went bankrupt. In 2010 more than a half million Florida homeowners lost their mortgaged homes through foreclosures.

Amendment 4's supporters could take at least small comfort in saying, "I told you so." After all, hadn't Florida's runaway growth created a glut of unsold housing and contributed to the state's economic mess?

A Turning Point

But many of Florida's top political leaders seemed unfazed by revelations that a glut of unsold houses contributed to the economic mess. Instead, they attacked the state's pillars of growth management and environmental protection. Leading the charge was Florida's new governor, Republican Rick Scott, an Illinois-born lawyer and businessman. He was the former CEO of a company he founded, the Columbia Hospital Corporation, which had been investigated by the federal government for fraud. Ultimately, the company had to pay $1.7 billion in fines to avoid criminal prosecution. The deal also required Scott to leave the company. He relocated to Naples, Florida; eight years later he ran for governor and was elected in 2010.

The Florida growth machine "needed 'bogeymen' and DCA conveniently fit the bill," writes Alan Farago, conservation chair of Friends of the Everglades. "It was a term actually applied to the agency by the ex-president of Associated Industries of Florida. DCA was his bogeyman." Such scapegoating helped deflect attention away from those who really helped cause the crisis.

By claiming the DCA stood in the way of job creation with too many rules and red tape, Scott and the legislature starved the agency of funds one year and gutted it the next. Remaining functions were farmed out to other agencies.

There were also budget cuts at DEP and the five water management districts. Cost cutting and the need to create jobs during the ongoing recession, said Scott, also made these firings necessary. Many top career scientists and administrators were sacked or demoted. Those who kept their jobs said they feared retribution—as many still do—if they didn't get on board the antiregulation bandwagon driven by the governor.

This brazen dismantling of Florida's attempt to manage runaway growth was cheered by the Florida Chamber of Commerce and the development community.

By 2016 signs of the Great Recession had faded, and the Florida growth machine shuddered back to life. Runaway sprawl soon returned.

However, many of Florida's most seasoned and dedicated professionals—scientists, engineers, administrators—who knew how to manage and protect Florida's water resources were long gone.

Demographers were predicting millions of people were once again moving to the Sunshine State. News of hurricanes, sea level rise, mass shootings, political scandals, dwindling resources, immigration woes, and education problems made little difference to the millions of humans from around the nation and the world determined to make Florida home.

Blind faith in growth was back. Realtors, bankers, and builders could almost smell prosperity in the new winds of change.

However, by this time voices could be heard from some within the business community, who, perhaps, discerned a disturbing truth they'd previously dismissed as the rantings of tree huggers.

It was, of course, the persistent question of the availability of water. Growth, after all, depended on access to plenty of water. It was time to do something, cried the hired orators of Big Development. More people were headed for Florida! Find new water sources! Soon!

To locate those sources, companies often seek help from site-location firms like the South Carolina–based McCallum Sweeney Corporation.

"When our firm scouts out possible relocation sites for corporate clients, we look at water from the tap to the source," explains Mark Sweeney, the company's CEO, "and we also look at availability, capacity, and the permitting of that capacity. The reliability of water is also important. If a company has to stop production because of a local government moratorium on supplies due to water shortage or quality problems, the company can quickly be in trouble."

The special challenge that Florida faces, in Sweeney's view, is the growing competition for water. "While it's true there's a lot of water in Florida," he says, "there's also a robust commercial sector, a growing population, and a small but potentially growing manufacturing sector—and all of them require water." Then, he adds as an afterthought, "There's also the Everglades, which needs water too."

In other words, the big question is: will there be enough water in Florida for everyone and everything in the future? Sweeney isn't so sure.

At many Florida water forums held in recent years, a bevy of industry-connected speakers echoed Sweeney's concern about water availability. Typically, they acknowledged the need to protect the Florida environment "that we all love." True, many admitted, Florida has growing water shortages. Just look at poor Orlando. The springs are sick, too. Way too much algae! Yuck! Yes, Lake Okeechobee is in big trouble. And there's sea level rise, too, but let's not talk about why.

Yet presenters generally remained sunny. That's because, they said, Florida can increase its population, expand suburbs and commercial areas, but still provide water for everyone without hurting the environment that we all love so much. "There's no water shortage," some insisted. "There's a water *management* problem."

Such talk, however, made some observers wonder if the speakers were suffering from some sort of "water supply denial" disorder. If water is finite, they ask, isn't there a point at which society must learn it can't consume forever?

The Israeli historian Yuval Noah Harari, in his 2016 best-selling history book, *Sapiens*, thinks many humans have trouble accepting the notion of limits. "Capitalism's belief in perpetual economic growth flies in the face of almost everything we know about the universe," he writes. "A society of wolves would be extremely foolish to believe that the supply of sheep would keep on growing indefinitely."

Is water any different from sheep? Is it really possible for the growth machine to plow forward and still provide water to all without serious consequences?

The answer from many of Florida's political leaders, water managers, scientists, and engineers is apparently and unequivocally, "yes!" They really do think it's doable. There'll always be plenty of water in Florida, they insist. All you have to do is to "grow the water pie."

7

Growing the Water Pie

THINK "WATER PIE" if you want to ponder things like a water planner. It's easy. Just picture a pie chart divided into slices. Each piece shows the fractional size of water used by a community's big water users.

But what happens when everyone—each stakeholder—wants more? How do you increase the size of that same pie, while keeping all portions constant? Figuring out such questions is the daily task for those men and women now being asked to really "grow the water pie" to meet future demand.

Most of them work at Florida's water management districts created in the 1970s by the Florida legislature. At the time, state lawmakers were under pressure from their constituents to do something to protect water sources. So the legislature set up six regional water management districts (WMDs)—later reduced to five. The WMDs' boundaries were drawn up to conform to those of natural water basins, rather than political ones.

The South Florida WMD is the largest of them all, reaching from Orlando to the Keys and encompassing the south and south-central parts of the state, including the Everglades.

Weighing in as the smallest district is rural Suwannee River WMD, occupying a fifteen-county region in North Florida often called the Big Bend area.

The Northwest Florida WMD in the state's panhandle stretches from the St. Marks River Basin in Leon and western Jefferson Counties to Escambia County's Perdido River.

The St. Johns River WMD follows the path of the St. Johns River, including counties making up Greater Orlando, which flows north to Jacksonville and beyond to the Georgia border.

Covering sixteen counties in southwest Florida is the Southwest WMD, often referred to by its nickname Swiftmud.

All five districts are charged with conserving and allocating water supplies, maintaining water quality, and providing flood protection and natural systems management. They are also taxing authorities that can levy property taxes and ad valorem taxes to raise money for water operations. The districts are answerable to the governor and the DEP. The governor also appoints members to serve on each district's governing board of directors. These individuals are volunteers and earn no salary.

Controversy often emerges when they must decide whether to approve requests for consumptive use permits (CUPs) that allow withdrawal of groundwater or surface water. Anyone receiving a permit is supposed to use the water only for the purpose for which it was granted; that purpose is also expected to be in the public interest.

Each governing board has a responsibility to allocate surface and groundwater resources in a sustainable fashion that is not appreciably harmful to natural environmental systems dependent on adequate water flows and levels. Florida's water guardians seldom deny permit requests, but they do sometimes reduce the amounts of water requested.

Over the years, many Floridians have praised the idea of local control at each of the districts. Who knows better, they said, how to deal with local water issues than the people who live there? Authority and power are diversified, too!

Today, though, that assumption is being called into question. For one thing, Florida's statewide water challenges are becoming so big and vexing that it makes sense for the districts to provide a bigger, unified response. It's also true the boundaries of the water basins don't seem quite so clear-cut as they did to Floridians in the 1970s. Engineers now realize that groundwater movement doesn't obey surface water boundaries. Another argument for recalibrating the district approach is that many big water users are extracting groundwater across district boundaries.

Small wonder, then, that four of the five WMDs have joined to produce *regional* water supply plans, in addition to their respective district plans.

The Central Florida Water Initiative (CFWI) is one of them. It's the product from an alliance of the St. Johns River, South Florida, and Southwest Florida WMDs, along with the DEP and the Florida Department of Agriculture and Consumer Services (DACS), regional public water supply utilities, and other stakeholders.

Backers of this huge combo-plan say it provides an integrated approach with a single set of defined rules, procedures, and policies for the three water districts whose boundaries overlap in Lake, Seminole, Orange, Polk, and Osceola Counties. This intersection is a growth hot spot where water demand is expected to increase from 800 million gallons per day to almost 1,100 million gallons per day in 2035.

The St. Johns River WMD, meanwhile, has partnered with the Suwannee River WMD to come up with the North Florida Regional Water Supply Plan. If all goes as expected, the plan's conservation projects are expected to reduce a projected 2035 water demand of 117 million gallons per day by 46 percent.

The regional plans are, however, just that, plans—though the CFWI did receive an extra dose of legitimacy when it was codified by the 2016 Florida Water Bill. Still, they are not self-implementing and aren't requirements, explains Claire E. Muirhead, regional water supply planning coordinator for the CFWI. Rather, each plan is a "guidance document; utilities have flexibility in what they want to implement. They can go outside the plan."

However, to receive water supply cost-share funding assistance from a district, a water supply project must be included in the plan, explains DEP's Janet Llewellyn. And if a WMD has already reviewed the project, a utility should have few permitting problems.

The DEP is optimistic that if all the water districts plans were to be implemented—and if the state legislature and other entities fund them—81 percent of Florida's new water supply needs will be satisfied.

Across the state, meanwhile, various utilities, municipalities, and regional water suppliers are charting their own way to water sufficiency. All want to diversify their water resources, if they can, to reduce the demand for groundwater. The trouble is the suite of options available to them is finite.

Water Pie Recipe

Conservation tops everyone's water pie recipe. It's the cheapest solution and sounds easiest. To channel the sentiment of Benjamin Franklin: A drop of water conserved is a drop earned. Reducing pollution, of course, would also seem an obvious place to begin. Removing contaminants not only protects habitat and ecosystems, it also renders an increase in the fresh water supply.

Just repairing and replacing aging, leaking pipes would go a long way in conserving water. According to a report from the Chicago-based Center for Neighborhood Technology, a nonprofit focused on sustainability, the United States wastes six billion gallons of water every day from faulty infrastructure.

Tiered water systems—paying higher rates the more you use, which already have been enacted by many utilities—are also listed on the water pie projects agenda. According to the EPA, if people worried more about how much they paid for water, they'd learn to conserve. This might mean paying more attention to leakage problems. Xeriscaping, or using natural vegetation that requires less water, can help too.

Inducement also plays a role. Since agriculture is responsible for 70 percent of all the fresh water used on earth, it may make sense to offer economic incentives to farmers to adopt conservation irrigation methods or water retention projects allowing them to "farm" water.

Governments may also want to use tax rebates to encourage homeowners, builders, farmers, private well owners, power plant operators, and business and industry groups to seek ways to conserve by choosing from an array of water-saving devices such as low-flow shower heads, dual-flush toilets (two handles, one for strong flush, the other weak), or toilets flushed by seawater or water from bathtubs and sinks.

Rainwater harvesting is also on planners' lists of things to do. This goal includes using more state-of-the-art stormwater systems that retain water instead of allowing so much of it to drain off.

Reservoirs can conserve water too, only it's not easy to store water in Florida. "Florida has two water problems," explains USF's Mark Rains. "Too much water and then too little. The water table is shallow, so when there's a downpour and the ground is soaked, there is no place for the water to go."

Water storage is a problem also, he says. "There aren't any huge canyons in Florida. Water is also inconveniently placed. Most of it is in North Florida, but most of the people are in the south."

Florida is flat. This condition makes runoff worse and storage a problem. Even if you transfer water from one county to the next, or one district to the next, where do you store the water when it gets there?

But it's also important to reduce evaporation loss in reservoirs. One novel approach to this problem is being pioneered in Los Angeles, which is experiencing the worst drought in its history. Here ninety-six million four-inch plastic "shade balls" were dumped into the city's 175-acre reservoir. Officials say the balls shade and cool the water, reducing evaporation between 85 and 90 percent. They also retard the growth of algae and bacteria.

Another way to retain water is to inject it into the ground, where it will be stored in human-built aquifer storage and recovery (ASR) areas until it's needed and hauled up to the surface for deployment. The drawback to this process, however, is that engineers may have trouble recapturing the stored water. Groundwater contamination from arsenic has also been a problem, though there are technological ways being developed to overcome it.

Recycling Water

"Water conservation is the first thing we must do," says Melissa Meeker, executive director at the Virginia-based WateReuse Association, a trade association dedicated to advancing water recycling, "but that has to be followed by actually increasing the available water supply, or growing our water pie—which means recycling water because it actually gives you new acre-feet locally."

To do this well, she suggests, requires rethinking and remembering that recycling of water isn't new. Nature's done it for billions of years. Modern water-treatment centers just do it faster.

Wastewater treatment is generally carried out by sending wastewater to treatment plants. There, it is filtered of solids and debris and chemically treated. In some plants it also undergoes advanced, or tertiary, treatment and then is pumped into an artificial wetland, farm, or public land where it is purified by plants before soaking back into the earth or being discharged to a river.

Chris Wildner, the lead treatment plant operator for the City of Ocala Water and Sewer Department knows this process well. As a young man, he got his first job in wastewater management in South Florida. In those days, he remembers, he shoveled sludge into containers to be hauled away. Today, forty-seven years later, he says he's amazed how much the "art of wastewater management has evolved, and helps you shepherd disgusting filthy water into something that's almost clean."

The Ocala plant, like so many American treatment facilities, relies on SCADA (Supervisory Control and Data Acquisition), a monitoring and controlling system used in industrial and manufacturing processes around the world. It uses sensors to track the wastewater every step of the way through the automated system.

Wildner watches monitors that graphically depict activities such as volume, flow rate, oxygen levels, pumping, filtering operations, and scrubbing action as they are happening. This frees him to think about problems and solutions, instead of wasting time walking through the complex taking readings as he once did.

The wastewater enters the plant, a grayish mass of leaves, trash, paper, plastic, rags, urine, feces, and who knows what else. From there it is filtered, sterilized, and generally spruced up enough to be sprayed safely upon a golf course in the nearby town of Belleview.

Water reuse isn't unique to Ocala. Nor is it new. Reused water, in fact, made its Florida debut during the 1960s on Southeast Farms in Tallahassee. A decade later it was being used for landscape irrigation in St. Petersburg. Today, sixty-three out of sixty-seven Florida counties reclaim wastewater, making Florida the national leader in this endeavor. Most of it is used to irrigate golf courses.

There are several advantages to producing reclaimed wastewater. For one thing, it reduces demand on surface and ground waters and helps recharge the aquifer. Using treated water also cuts down on the cost of developing new water sources. For farmers, it offers a steady supply of water. Growers also may be able to reduce their use of fertilizers, because treated water often contains nutrient residue. Sending wastewater to treatment plants, of course, also helps reduce pollution.

Drawbacks exist, too. Reuse water can be applied only to crops that must be peeled or cooked to eat. Although recycling enthusiasts like to say that science can take everything out of water except the hydrogen and oxygen, some studies show that reclaimed water may at times be too salty for certain crops. At times, tiny trace amounts of chemicals or nutrients still exist in treated water that trigger microbial growth that's harmful to plants, though good water-quality monitoring can minimize this risk.

One group of bioactive chemicals raising concern among scientists is pharmaceuticals. Every year, thousands of tons of antibiotics, mood stabilizers, sex hormones, and an array of active ingredients in personal care products are dumped into toilets and flushed into wastewater systems. Many of these substances also find their way into supplies of fresh water. So far, they have appeared only in tiny amounts measured in parts per billion or trillion, which utilities say makes them safe.

Still, more than a thousand pharmaceuticals have infiltrated the natural world, and researchers don't yet fully understand the impact they may have on living things. Many more drugs are expected to infiltrate the water supply in the near future, as humans live longer and require more medical care. So far, the EPA doesn't require utilities to test for them, though some do so on their own.

Meanwhile, there is no doubt that nutrients in recycled water can cause problems. For instance, Wakulla Springs has gone through "ecological collapse," says springs expert Jim Stevenson, in large part because of nitrate in sewage effluent that was sprayed onto a nearby two-thousand-acre spray field by the City of Tallahassee. New advanced treatment, however, has decreased nitrate loading by more than 66 percent in recent years.

Tallahassee isn't the only city with water recycling woes. During the summer of 2017, Titusville officials had to warn residents that they couldn't use reclaimed water because it wasn't up to DEP standards.

"The trouble Florida farmers have with recycled water is that they are always worried if the water will pass federal food safety standards," says Charles Shinn, director of Government and Community Affairs at the Florida Farm Bureau Federation. "You have groups like Walmart, Safeway, and Target that are dictating to farmers exactly what water resource they must use on their crops." Reuse water might not be good enough.

Reliability proved to be another problem. In the 1980s the City of Orlando and Orange County teamed up to create Water Conserv II, a project that sent treated wastewater back into the ground through "rapid infiltration basins" to recharge the aquifer and also to provide irrigation for agriculture. It was the first reuse project of its kind permitted by the DEP.

"At first, citrus farmers were glad to get it," says Shinn. "But eventually 'purple pipe'—or recycled—water became popular, and Orlando utilities decided they could make more money by selling it to golf courses, and so they priced agriculture out of the reuse market, except for a few nurseries."

Treating wastewater also requires lots of energy. Too many treatment plants, however, are inefficient; in fact, according to the U.S. Department of Energy (DOE), wastewater treatment plants in the United States use five times more energy than they should. That's a problem, says the DOE, because wastewater operations, "typically the largest energy expense in a community," are expected to increase "by up to 20% in the coming decades due to more stringent water quality standards and growing water demand based on population growth."

The good news is that wastewater facility retrofits can produce up to 50 percent in energy savings, say energy experts, which would help lower the cost of cleaning our increasingly dirty water. Ironically, this process requires water to produce the electricity needed to power the retrofits.

Demand for electricity, thus, should rank high on the long list of worrisome things to keep Florida's water pie planners up late at night. That's because electricity and water share a sort of symbiotic relationship.

The Union of Concerned Scientists reports that 65 percent of U.S. electricity is produced by thermoelectric power generators that are cooled by water. These plants also require lots of water in the form of steam to operate turbines to produce electricity. However, to boil water to make steam, you need a fuel, such as natural gas, oil, or uranium. Drilling is required to free these resources from the earth, a process that, once again, depends on water.

Electricity also is indispensable for pumping and transporting water. Pollution control technologies at the power plants also require water.

So it's no small wonder scientists like to say, "Conserve energy, conserve water."

Getting Clear Eyed

There's also the issue of water's intrinsic worth. "I think the biggest thing we can do is address our public perception issues and get people to actually value water for what it's really worth," says Melissa Meeker, who is also a former executive director of the South Florida Water Management District and past deputy secretary of the DEP. "Right now, we undervalue water. That's especially easy to do in South Florida, where it rains a lot and our focus is to get rid of the water as quickly as possible. Too many of us don't think about water or what it is needed for—including life itself. But we need to step back and think of water holistically."

Meeker believes if people really undertake these reevaluations, they will understand the need for reuse. "As our water challenges become more apparent, we'll have no choice," she says, but to reconsider water in an integrated way. "When you say reuse, you could be referring to domestic wastewater, stormwater, rain catchment water, or industrial processed water—all of which can be treated and used over and over again."

That might even include drinking potable reuse water, as did Jack Black and the beer drinkers who imbibed a brew made from Toilet to Tap water provided by WateReuse. Meeker agrees there's a yuck factor to overcome first. But that's been done already in California, she says, where the "toilet to tap" expression came from.

It took a while to achieve, though. "Fifteen years ago, the proposal to use potable reuse in that state went down in flames," Meeker adds.

Today, though, because of prolonged drought, combined with educational outreach efforts promoting potable reuse water, attitudes have changed. For example, in Orange County, California, new state-of-the-art advanced water treatment technology is turning wastewater—that would otherwise be wasted to the Pacific Ocean—into water that officials believe is pure and safe enough to supplement the county drinking water supply.

Finding New Water?

Water pie managers like to think they have a suite of options. One of them is to find "new" water. Typically, this means tapping a surface water body or an aquifer that you weren't using previously. The CFWI, for instance, sees the St. Johns River as a new source.

Some Florida communities, like Ocala, are sinking deep exploratory wells down to the Lower Floridan Aquifer (LFA) that lies beneath the Upper Floridan Aquifer (UFA)—the one currently being pumped. Scientists disagree, however, whether the two are really separated by a confining layer, or if there's just one aquifer.

Extracting super deep water may not be worthwhile. Engineers say it's riskier to drill in the LFA, where less is known about well capacity. Treating any water pumped up from the extreme deep aquifer also may prove too expensive.

There is even interest in freshwater sources located in the sea. As many scuba divers and anglers will attest, numerous freshwater springs lie off Florida's coasts, many of them in estuaries or bays. One of the better known is the Crescent Beach Spring, situated 2.5 miles out in the Atlantic Ocean off St John's County. According to the U.S. Geological Survey, this submarine spring's "discharge is large" and "greatly exceeds" the volume of water produced by first-magnitude springs.

Currently, the Florida Geological Society is sponsoring a search for other freshwater springs off the coast of Florida. Though it's too soon to speculate, some experts suggest that one day it may be possible to pipe this offshore fresh water to coastal cities.

Even so, it's unlikely anybody will find a problem-free fountain of fresh water. According to a CFWI staff-written response e-mailed to this writer, district studies indicate this offshore water would need costly treatment. There's also "cost and potential environmental issues associated with transporting this source onto the shore."

The CFWI staff also believe the groundwater for these offshore sources is typically from the same aquifer as in-shore sources. Using them, they say, would not "necessarily resolve issues associated with groundwater withdrawals and possible impacts to water resources on shore."

Offshore springs may not play a big role in future fresh water supplies. But seawater could possibly play a starring role.

Getting the Salt Out

The process of extracting fresh, or sweet, water from brackish or salt water (desalting or desalination) isn't a new idea. In fact, it's been well under way in Florida since the 1970s. Nor is removing salt from water a grand innovation. In his work *Meteorologica*, Aristotle, the fourth-century BC Greek philosopher and scientist, observed, "Salt water when it turns into vapour becomes sweet and the vapour does not form salt water again when it condenses."

Seventeenth-century seafarers reportedly distilled water. It's easy to do. Simply boil brackish or seawater until it evaporates into steam, leaving behind a residue of salt. Then condense the steam into fresh water in cooled pipes.

In the United States, large-scale distillation began in the 1930s. No modern water utility, however, uses the evaporation method much anymore. More efficient methods have taken over, the most common being reverse osmosis (RO). This process involves, say, brackish water from an estuary that is pumped into a desalination plant, where it goes through a filtering process to screen out the trash and organic solids. It's also treated with chemicals to knock out microbes and other things deleterious to humans, such as algae. Then the water is subjected to tremendous pressure that forces water molecules through special filters made of membranes with holes so ultra-tiny that they catch salt and let the water particles through.

RO's origins can be traced back to the 1950s, when researchers developed special polymer films that could separate salt from water. The great Aristotle, perhaps, also may have seen RO coming. "Water," he wrote, "which gets through the wax walls is fresh, for the earthy substance whose admixture caused the saltness [sic] is separated off as though by a filter."

Once the water has passed through the RO filters, it is still not ready for service. It's treated again, this time to restore the water's correct chemical balance to make sure it won't react with the metal pipes. The added chemicals will also reflavor the water so that it tastes right. Next, it's time for some more filtering, and finally the desalinized water is sent off to the freshwater supply lines.

Today, Florida leads the country in the use of desalination water in both the number of plants and gallons produced. There are more than 140 facilities statewide, most of which convert brackish surface or groundwater to sweet water. Seawater isn't used for the process as much.

Florida's king of desalination, which does use seawater, is the regional utility Tampa Bay Water. Until recently, its seawater desalination plant was not only the biggest facility of its kind in Florida, but also in the nation. That distinction recently passed over to a newer facility in Carlsbad, California.

Nonetheless, Tampa Bay's Apollo Beach plant is still a big deal. It is part of a $1-billion urban water supply system that interconnects surface, ground, and desalination water. This three-legged structure is the only one of its kind in the United States. "It's like a financial portfolio," explains Ken Herd, a plant engineer and construction and contract section leader. He is also one of the engineers who created the facility. "It's one of the most diverse systems in the world," he tells a group of visitors in the lobby of the Tampa plant. "The system is drought resilient."

To do all this, engineers came up with a new three-part system: a surface-water treatment plant that produces about 120 million gallons per day, a 15.5-billion-gallon reservoir to supply water when it's needed, and a pumping system that is permitted to pump 120 million gallons per day of groundwater from wells.

"We don't have to run the desalination plant at all during the summer months, because there's enough rainfall to recharge the surface and groundwaters," says Herd. So they make potable water when rain is rare. And if the utility needs more water, it can always use its reservoir or its groundwater supply. Each component serves as a backup supply for the other two.

When the Tampa plant does "make" fresh water, it uses reverse osmosis. The process, however, is expensive and requires a lot of energy. One way to reduce costs is to increase efficiency by building the RO plant next to a power plant. Engineers call it "co-locating."

That's what Tampa Bay Water did. Its huge RO system sits next to the Big Bend Power Plant at Apollo Beach, a few miles south of Tampa. Brackish water used to cool towers at the power plant is afterward piped over to the Tampa water works for an RO treatment.

What to do with the leftover highly concentrated salt water after the desalination process is complete isn't as big a problem as some people think. Herd explains that plant managers pipe the leftover briny soup back over to the power plant, where it is blended and diluted with less salty water and pumped back into the bay.

Care is taken to make sure that salinity levels don't harm marine life in the estuary, says Herd. What about sea creatures getting caught up in the plant's hardware? Herd seems puzzled by the question. There's no huge pump that sucks in the salty water. The system is gravity fed. Screens keep the fish out. "Mollusks may get trapped, now and then," Herd admits. "But that's about it." If so, it may be because the water Tampa Bay uses comes from the Big Bend power plant that does take a big toll on aquatic life when it intakes and filters seawater for its cooling towers.

Saving It Up for a Sunny Day

Drive a half hour east of the Apollo Beach plant, and you arrive at another important component of the Tampa Bay water supply portfolio: the 15.5-billion-gallon C. W. Bill Young Regional Reservoir. Here, in a remote part of Central Florida near a rural area called Ft. Lonesome, looms a huge eighty-foot-high structure that looks like a swimming pool for giants. It is two miles long by one mile wide, or eleven hundred acres. The reservoir's depth ranges from four to eighty feet. It takes two riding lawnmowers two weeks to cut the grass growing on the embankment of the facility.

The big pool holds surface water that is pumped in from the Tampa Bypass Canal and the Hillsborough and Alafia Rivers. After a good summer downpour, water is stored in the reservoir; thus, Tampa Bay Water can literally save the rainwater for a sunny day. When the weather becomes dry and groundwater levels drop, the reservoir discharges some of its water. The desalination plant can crank up production, too, if necessary.

The C. W. Bill Young Reservoir first went into service in 2005, but it had to be shut down in 2012 when cracks appeared inside the structure. The engineers remodeled the facility and put it back on line in August 2014.

One of them is Rich Menzies, who has worked on the reservoir project since day one. He's soft spoken, courteous, and eager to take people on a tour in his white SUV around the perimeter at the top of the reservoir. It's easy to see that, like Ken Herd, he is proud of the work he and his colleagues have done. "I liked to play with tractors and dump trucks as a boy," he reflects, "and I still do. That's part of why I love what I'm doing here."

Because the surface water collected at the reservoir could remain there for months at a time, compressed air is sent through seven forty-foot-tall aeration towers placed in the reservoir, along with air diffusers on the bottom of the impoundment to increase the oxygen levels to prevent algae from growing and to keep the water tasty. Another problem is evaporation. When this happens, fluorides, released by phosphate rock mining and brought to the reservoir by the Alafia River, become concentrated. The utility prevents these fluoride amounts from reaching toxic levels by blending treated desalted seawater and treated surface water "as final, finished waters at our regional site," says Tampa Bay Water spokesperson Christine Owen.

That oxygen-rich reservoir water also supports the fish that arrive with the river and canal water. Gators live in the reservoir, too, only they find an alternative way to get in. "They come up out of the wetlands below," says Menzies, gesturing to a swampy area down below and outside the reservoir. "They can scale that eight-foot metal fence and crawl up the embankment and get into the reservoir."

A few gators, in fact, were lazing not far from Menzies as he continued. "After a while, we began to notice that many of them were missing limbs. At first, we thought they were cannibalizing one another."

That wasn't the case, however. Florida Fish and Wildlife Conservation Commission experts who were called in to investigate decided that during mating season the gators became territorial and attacked one another. There was nowhere for them to retreat to because even though the reptiles could get into the reservoir, they couldn't get out. "The concrete seawall prevents the gators from exiting the reservoir because of its curved edge facing the water," Menzies explains. "There are gated openings in the wall that are now left open during mating season to allow the gators to exit."

So when biological imperative hits, the reptiles can sally forth to reproduce in the wetlands. You can't help but compare the enlightened care provided at the reservoir with the kind of treatment Florida crews back in the dredge-and-fill days would have meted out to any unwanted gator intruding into their workspace.

Gators are the biggest creatures in the reservoir right now. But they might have had some extra company if all had gone to plan, because the reservoir was supposed to have doubled as a recreation area for people. The events of 9/11, however, mothballed that idea, and today security of the premises has been beefed up to keep most people out.

Tampa Bay's multiprong approach to growing the water pie seems to be working, at least for now. Utility officials believe their integrated system will meet the area's needs for the next fifteen years. Additional drinking water might be needed after that, but how that extra supply will be found hasn't been determined yet. One possible source is a new seawater desalination facility co-located with the Anclote Power Plant on the Gulf of Mexico. The City of Tampa is also hoping to use potable reuse water.

Can They Do It?

Water planners use computer modeling, tons of data, and assessments from a slew of experts to help them gaze into the next twenty years. But they don't have crystal balls. As always when making long-range plans, no one can anticipate all the obstacles and unexpected consequences.

Now and then, however, thoughtful planners do anticipate future impacts, but convincing others to see it, too, isn't easy. This happened to Margaret Spontak, a former St. Johns River WMD water planner.

"In the late 1990s, serving as director of policy and planning at the district," she says, "I questioned Barbara Vergara, head of Water Supply Planning about long-term projections for the water supply in Marion County. She told me that it would never have a water supply problem. I asked her if the modelers had factored in The Villages, a development south of Marion County that had fewer than 8,000 residents then, but was projected to grow to 50,000 residents. She said no. Within a year, additional modeling was done, and the Water Resource Caution Area Map reappeared with a caution area in south Marion County. The Villages now has a population of approximately 160,000. Despite that designation and unprecedented growth, permits have continued to be issued as normal. Silver Springs has lost at least 32 percent of its historic flow. Lake levels throughout south Marion County are disastrously low."

Alternative Water Truths

There may also be times when a water management district staff sees danger ahead, and that truth is prevented from coming out. That's the verdict of hydrologist Jim Gross, who worked as senior project manager, technical program manager, and assistant division director for the St. Johns River WMD.

"The governing board was not allowed to see a 2010 water supply plan which clearly concluded that there was a lack of sustainability in the district," he says.

In 2013, again word came down from Tallahassee, according to Gross, that the district's water supply plan was again not allowed to go to the governing board.

"At the time," he says, "I was at the center of regional water supply planning. I talked to managers, scientists, and engineers, and from what they all said, it was clear to everyone that there were explicit directions from management that Tallahassee wanted them to put this plan on the shelf until later."

Gross makes it clear he disapproves of these decisions. He's also bothered by something else: an internal struggle within his district in 2014, when he and district staff recommended against a permit request from Sleepy Creek Lands (the former Adena Springs Ranch) in Marion County.

Grass-Fed Cattle Controversy

At first, Frank Stronach, a Canadian billionaire and Sleepy Creek Lands' prime mover, wanted a permit allowing him to pump up to 15.2 million gallons per day from the aquifer, an amount roughly twice that of what the fifty-eight thousand residents of nearby Ocala consumed every day.

Stronach's request for such a big slice of the water pie was so he could irrigate pastures of an unfolding and potentially huge grass-fed cattle enterprise. At the time, the market for the leaner beef looked promising.

Fast and furious came public opposition. Several environmental groups and individuals took legal action.

Jim Gross and his staff, meanwhile, were using a groundwater computer simulation program called the North Central Florida Groundwater Flow Model (NCF). After running dozens and dozens of NCF simulations to find ways to recover the lost flow at Silver Springs, they decided the "results indicated conclusively that" groundwater pumping would need to be significantly reduced. Thus, they also concluded, the extra water extraction needed for Stronach's project would hurt Silver Springs and the Silver River. That's why in 2014 the district issued a notice of intent to deny the Adena permit.

A year later, the ax fell on Gross's public-service career. "I was the first person to get canned after the second inauguration of Rick Scott. The district fired a number of people, including me, who were directly associated with making the district's decision not to issue the permit to Stronach. Even the executive director was fired."

There was another change. The NCF, says Gross, "was then thrown in the trash can and replaced with another program"—a newly revised version of Swiftmud's Northern District Groundwater Flow Model. The newly reconstituted St. Johns River WMD staff used the new model—one that combined data from Silver and Rainbow water basins—and concluded there was plenty of water for pumping after all, even though the decline of both springs was hard to miss.

Critics charged that backroom politics had corrupted the process and that the new model showed flows and levels that didn't exist to make it look like there is more water than really existed.

The district, however, responded it was simply using a newer and better model that brought new results.

In the summer of 2017, after a welter of protests and legal challenges, Stronach finally received a water permit, but it was far less than what he wanted at first. Even so, his opponents believed the water district had given him too much and by doing so would add to the ongoing degradation of Silver Springs and the Ocklawaha and St. Johns Rivers.

Aftermath

Things haven't gotten better since then, Gross says, here or elsewhere in Florida. "Today, we have crossed the line of sustainability," he asserts, "and the districts are trying to manage water resources by putting at odds one interest against another. Scientists now have been put in conflict with those who don't like the answers they're given by scientists about the water supply. These sustainability problems are apparent on the west coast of California, the Great Plains, parts of the Northeast, and now even in Florida. We're straining these resources."

Today, Gross, still upset about his ouster, is the executive director of Florida Defenders of the Environment and a professor of geology at Santa Fe College.

He's also a big critic of the North Florida regional water plan. "It's too little, too late. The plan calls for nothing that would lead to solving the crisis. There's no real acknowledgment that there is a water crisis."

The authors of the CFWI water plan, however, do at least express concern. Read the plan's executive summary, and you detect a sense of urgency: "[T]raditional groundwater resources alone cannot meet future water demands or currently permitted allocations without resulting in unacceptable impacts to water resources and related natural systems . . . In some areas, utilization of traditional groundwater is near, has already reached, and in some areas has exceeded the sustainable limits. Adverse impacts from withdrawals are already occurring in several areas."

Water management officials also respond they are protecting water sources, because state law requires them to develop minimum flow and levels (MFLs) of a water body before extracting water from the recharge area. Level, or elevation, refers to the depth of the water body. Flow is used to measure the volume of water that moves a certain distance. Minimum flows are set for Florida's rivers, streams, springs, and estuaries. Lakes, wetlands, and aquifers get minimum levels.

Miffed over MFLs

The history of MFLs in Florida, however, is controversial. Staff scientists at the districts often spend years doing fieldwork to establish them. Surveys are done. Water elevations are established. Gauges measure water flow. Various habitats are evaluated. Scientists go underwater. They tromp around in wetlands and flood plains, taking soil samples and examining the vegetation. They are mindful of things such as fish passages and the "regime" of a natural flow. Increasingly, they must contend with problems of algal growth and submerged aquatic vegetative growth that proliferate so much that they restrict flow.

After all their fieldwork is done, data is collected and crunched by computer modeling to determine MFLs. The results are peer-reviewed and opened up to public commentary. The water management district governing boards then vote whether to approve what their scientists have recommended.

Attend WMD public hearings, and you'll hear staff scientists insist they have nothing but good science to offer. Nonetheless, there is a lot of public unease over the setting of MFLs. For instance, many nonagency scientists and activists often publicly question how the data was collected. Were there multiple readings at the sample sites, or a few? Over what time period were these studies made? After all, if you sample too short a span of time, you may see only a snapshot of a river's conditions. Should you go far back in time—say a half century—and include historical highs and lows to get more a climatic view? But if you do, you run a different risk. Suppose the study period dates back before the year 2000—that is, before the arrival of frenzied groundwater overpumping. In that case, the study will include a longer period of time when water conditions looked much better than they do now. Critics of this kind of sampling say it would be like trying to figure out whether a sixty-year-old house has a current termite problem by looking back at all annual pest-control reports. It doesn't matter that the house was bug free when it was new. What matters is now.

Arguments also often erupt over the use of computer models to determine MFLs. How computers are calibrated is one thing sure to provoke disagreement among scientists.

How Much Harm Is Okay?

Another point of contention invariably pops up at public meetings and administrative law proceedings over water rule Section 373.042(1), F.S., which states, "The minimum water level [or flow] shall be the level . . . at which further withdrawals would be significantly harmful to the water resources of the area."

The words "significantly harmful" caused a big stir at a Swiftmud public meeting at its Brooksville headquarters on March 28, 2017. Here, despite the overwhelming objections of the clear majority of one hundred audience members—mostly kayakers, ranchers, scientists, and activists—the water district governing board adopted a new MFL limit for the Rainbow River, one that will allow an extra 5 percent of the flow of the already troubled river to be extracted from its recharge area.

Earlier, during the public comment period, staff scientists were questioned about how they had come up with the 5 percent figure. A spokesman responded they'd used an international standard that allows harm resulting from increased water extraction to be inflicted as long as it does not exceed hurting 15 percent of the local ecology. Thus, the harm caused by allowing Rainbow River to be lowered by an additional 5 percent reduction in an MFL would be okay because it would not exceed the 15 percent threshold and cause "significant harm."

But what is significant harm? How is it different from everyday harm? Ask ten scientists, water officials, or attorneys that question, and you're likely to get just as many opinions.

Even Florida water law is somewhat fuzzy on this matter. Minimum flows and water levels, say the statutes, are those that if exceeded "would be significantly harmful to the water resources or ecology of the area."

Who decides those calculations? It's the DEP or the governing board of each water management district—political appointees, that is—who presumably will be "using the best information available."

Not all that information is based on science, however. Here's what administrative law judge J. Stephen Menton had to say in the 1997 case between Charlotte County and the Southwest Florida Water Management District: "The establishment of minimum flows and levels does not have to be based on precise historical averages. The statute seeks to prevent 'significant' harm to the water resources. Preventing any and all measurable impact to the water resources is not the stated legislative goal and some impact is an unavoidable element of achieving beneficial use of the water resources for human activity. Thus, the establishment of MFLs is highly infused with policy considerations and requires a balancing of societal interest in order to decide what impacts are significant."

Societal interest? Who, then, decides "society's" interest in allowing a river flow to be lowered another 5 percent? The administrative judge? The water districts' governing boards? The governor? Developers?

If social policy—and not just best science practices, as many state officials are fond of saying—can be used to determine Florida's water quantity and quality issues, it's no wonder that more people want to have a say-so over how to divvy up the water pie.

In fact, at that Brooksville meeting, a few of the mostly angry and discouraged people in the audience accused the governing board of adopting the MFL change to provide more water for development, even if that meant hurting the Rainbow River—a charge one board member hotly and indignantly denied.

The acrimonious exchange that spring day did illustrate one important fact: people will become agitated, passionate, and perhaps even militant if they don't trust their water stewards.

This lesson gives rise to the next logical question: Who owns the water anyway?

A tranquil setting on the Withlacoochee River (or Crooked River), which begins its journey in Central Florida's Green Swamp and then flows 141 miles northwest to the Gulf of Mexico near Yankeetown.

In October 2014 the Expedition 41 crew aboard the International Space Station took this photograph, which reveals the extent of human settlement in Florida and that state's vulnerability to sea level rise.

White Springs as it appeared in 2005, now dried up. The largest single cause of the loss of water is the diversion of water to a nearby phosphate mining facility.

People at the Spring House bathhouse at White Springs, Florida, in 1914.

An alligator killed by pollution in Palm Beach, Florida, in 1969.

On September, 13, 2018, boaters at an Orange Springs boat ramp in Marion County were greeted by an outbreak of two invasive plants—water lettuce and water hyacinth.

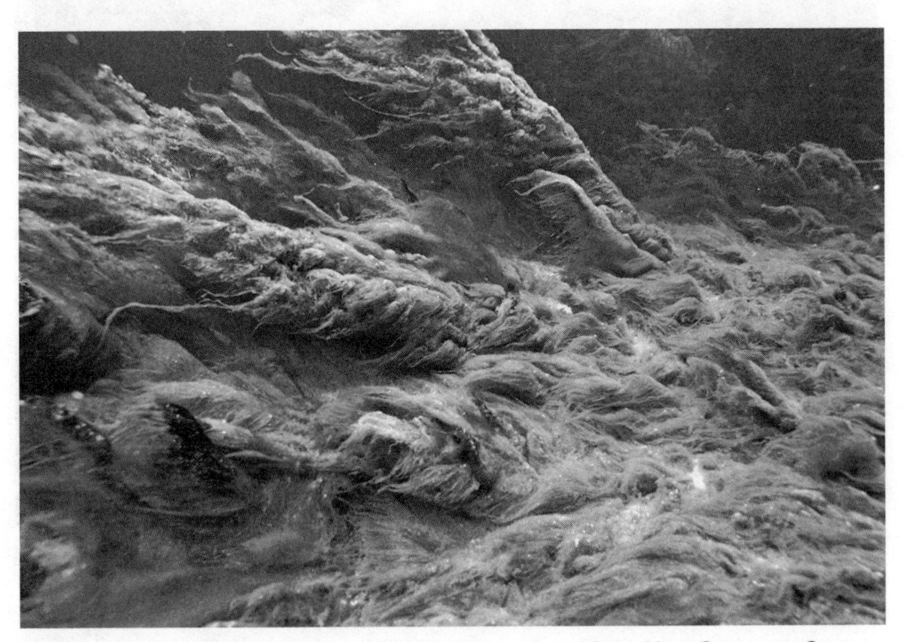

Runaway algae smothering natural underwater plant life at Hart Spring in October 2007.

Mermaid Bonita Colson poses underwater at Weeki Wachee Springs in 1960, when Florida's springs were still crystal clear.

In March 2018 Silver Springs, now a state park, provided a small fleet of glass-bottom boats that allowed visitors to view the springs, though their clarity and strength had been substantially diminished in recent decades.

A manatee at Wakulla Springs.

In January 2017 Florida springs scientist Robert Knight spoke to protestors gathered outside the St. Johns River Water Management District headquarters. To his right is St. Johns Riverkeeper Lisa Rinaman.

This aerial view reveals widespread algae pollution at Kings Bay at Crystal River.

Left: A portrait of industrialist Hamilton Disston, who in 1881 saved Florida from bankruptcy by purchasing 6,250 square miles of land from Orlando to south of Lake Okeechobee for $1 million and launched an era of dredging to rid the state of "excess water."

Below: The dredge *Reclaimer* at work on the Everglades drainage project sometime during the 1920s.

One of several communities precariously situated within a few hundred feet of the top of Lake Okeechobee's Herbert Hoover Dike embankment in March 2017.

The forest-leveling machine known as the Crusher-crawler was used by the Army Corps of Engineers to clear forests for the construction of the Cross-Florida Barge Canal.

A bird's-eye view of a section of Everglades City after the arrival of the deadly hurricane of September 1926.

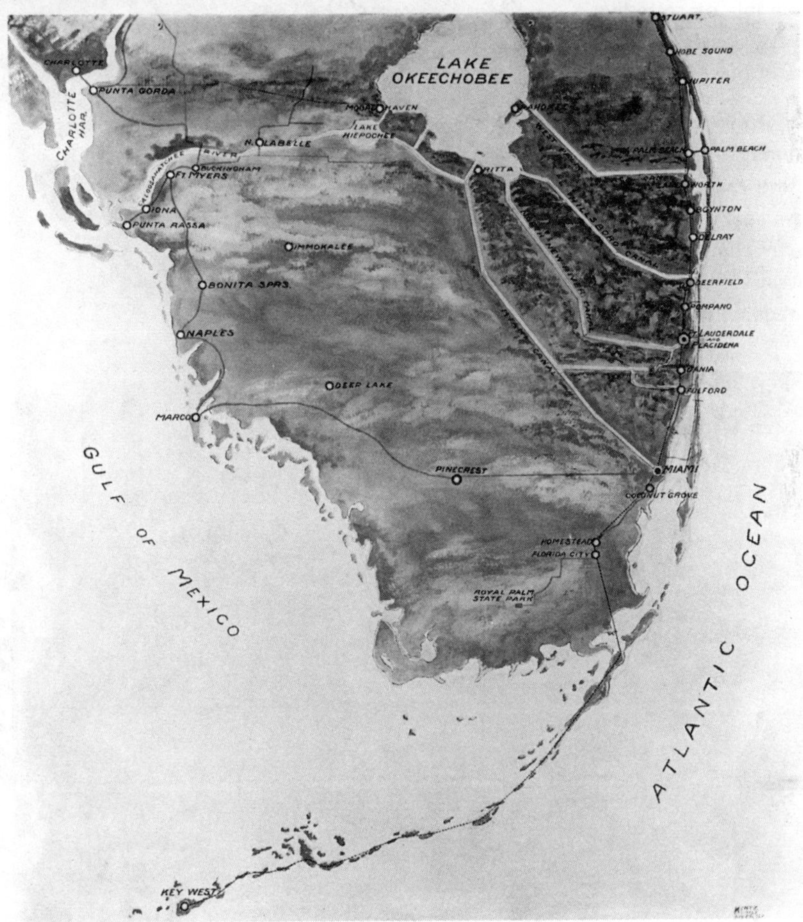

A 1924 rendering showing the location of canals in South Florida created to drain the Everglades.

Right: Portrait of Dr. John Gorrie of Apalachicola, whose pioneering work in air-conditioning to comfort his malaria patients also made Florida a more hospitable place to live, thereby attracting millions of newcomers.

Below: In 1992 Tallahassee demonstrators, representing various sectors of the real estate and construction industries, protest Florida's Growth Management Act.

An aerial view of The Villages, a sprawling retirement community and one of the fastest-growing U.S. cities, which spreads from Sumter County to neighboring Lake and Marion Counties.

The 15.5-billion-gallon C. W. Bill Young Regional Reservoir, a component of Tampa Bay Water Supply system.

Inside the Tampa Saltwater Desalination Plant.

View from above of the St. Johns River, which Central Florida water supply managers consider a source of fresh water.

Left: Senator Reubin Askew, Democratic nominee for governor, making a campaign speech in Highlands County, Florida, in 1979. Once elected governor, he successfully encouraged passage of several landmark environmental protection laws.

Below: This map shows the boundaries of Florida's five Water Management Districts, established to conform to the state's natural hydrological boundaries.

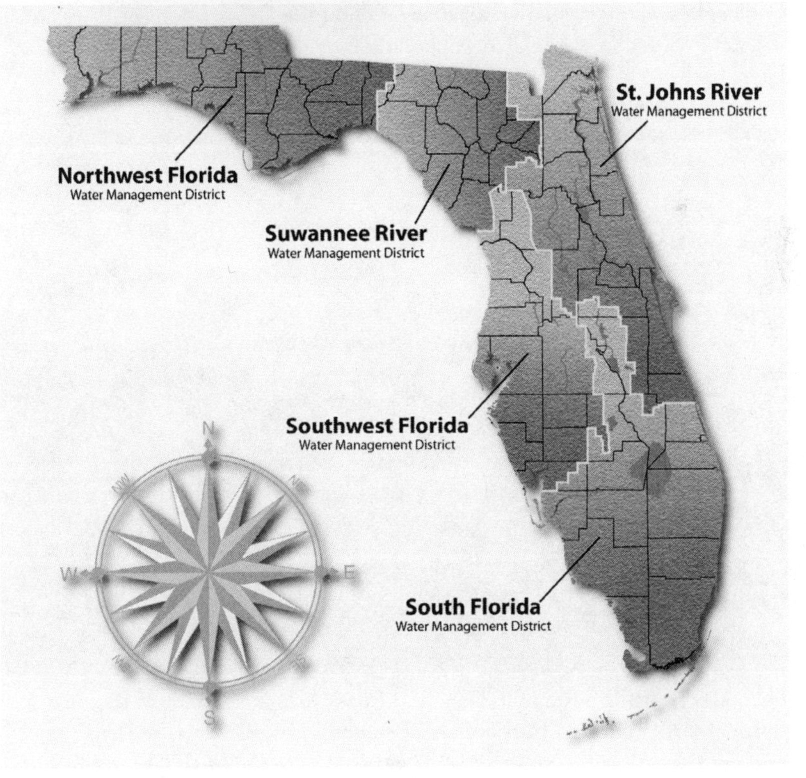

Governor Rick Scott speaking at a Tea Party rally in Tallahassee in May 2012. By this time, Scott had significantly reduced budgets and personnel in Florida's water management districts and the Department of Environmental Protection, while also weakening Florida's growth management laws.

In January 2017 the Rodman Dam and Reservoir, a remnant of the now defunct Cross-Florida Barge Canal Project on the Ocklawaha River, remained the epicenter of a regional water dispute in North Central Florida.

The fate of the picturesque town of Apalachicola, along with that of the Apalachicola River and Apalachicola Bay, depends on how Alabama, Florida, and Georgia divvy up a water resource common to all three states.

An aerial view of a sinkhole at Winter Park that formed in 1981, swallowing a house and causing an international sensation. Such events can be triggered by drought and groundwater pumping.

A portrait of Marjory Stoneman Douglas, journalist, suffragette, social activist, author, and environmental activist. Nicknamed the "Grand Dame" of the Everglades, she received many awards throughout her life, including the Presidential Medal of Honor. In 1990 Broward County, Florida, christened its new high school in honor of the one-hundred-year-old Douglas.

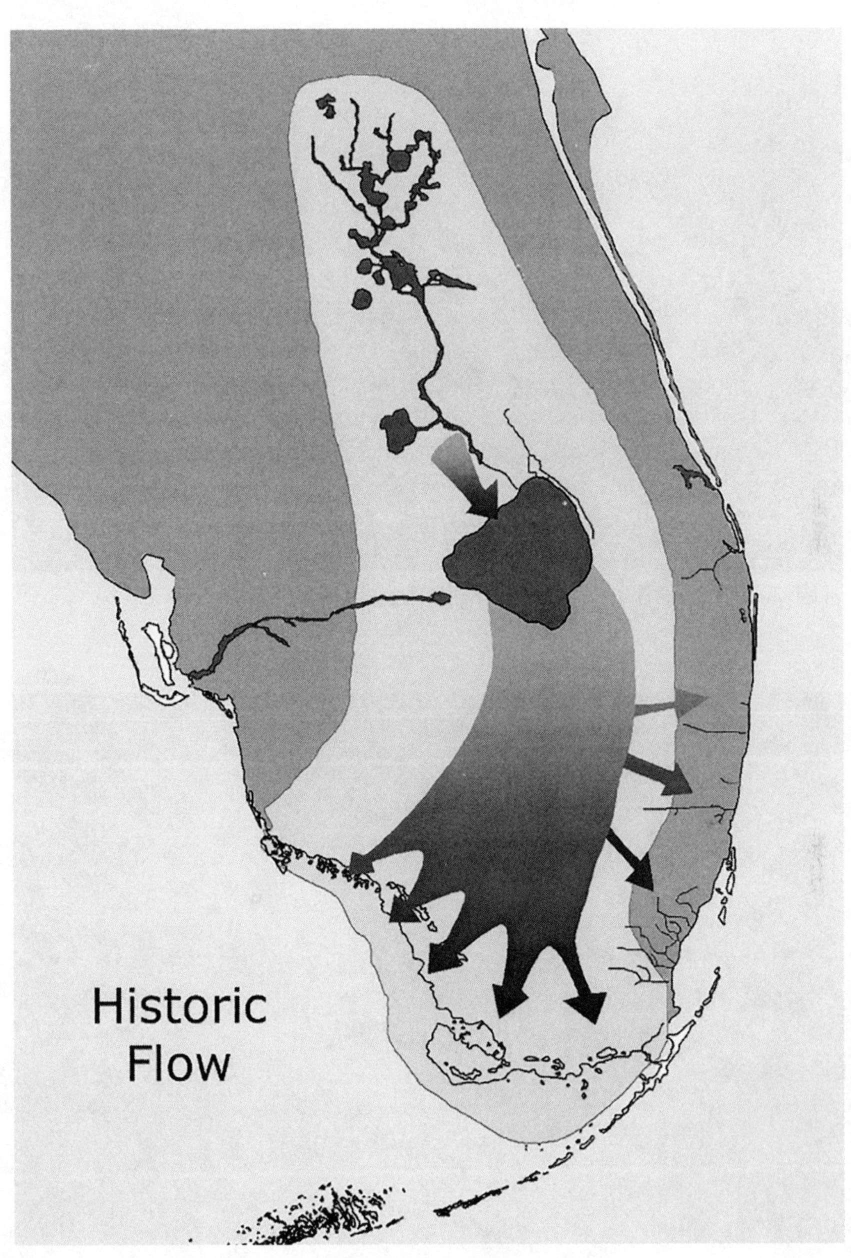

Map showing the historic flow of the Everglades.

Aerial view of Lane River in the Everglades National Park.

The Florida panther is one of several endangered species in Florida imperiled by relentless development. The Everglades is one of its last refuges.

Visitors watch an alligator crossing the Anhinga Trail at the Everglades National Park.

People wading on flooded Lenox Avenue in the city of Miami Beach, during the aftermath of Hurricane Sandy on October 28, 2012. In recent years the city has spent tens of millions of dollars trying to protect the community from natural flooding events made worse by sea level rise.

Sea level rise resiliency project at Cedar Key in October 2017.

The 15,500-acre Storm Water Treatment Area 2 in southwestern Palm Beach County treats water before it is released to Everglades Water Conservation Area 2.

Two views of Lake Howard Nature Park, one of three city parks that receive and treat stormwater in a natural way in Winter Haven. They are part of the city's "green infrastructure" approach with planning assistance from water consultant Tom Singleton.

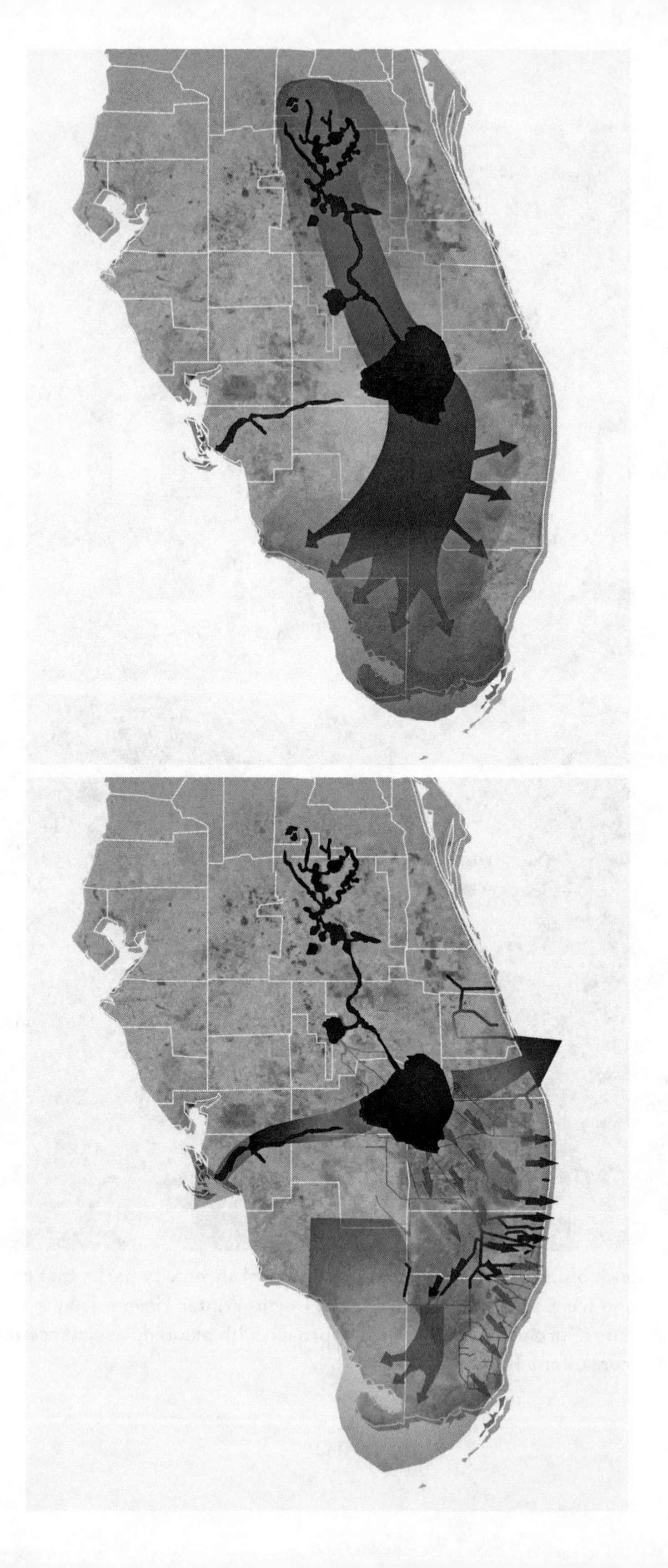

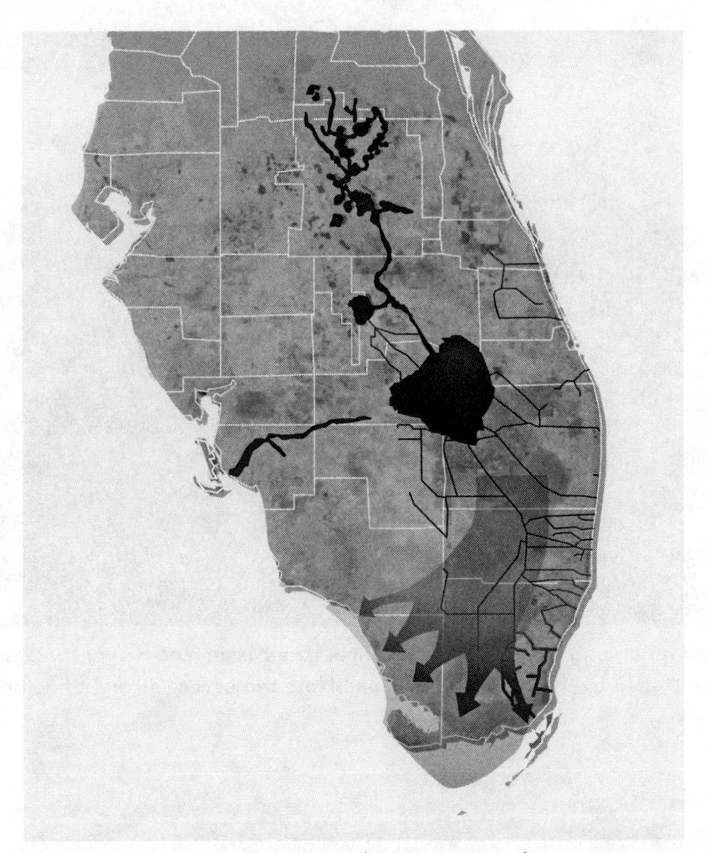

The first of these three images (*facing page, top*) shows the historic water flow in the Everglades, the second (*facing page, bottom*) the current flow, and the third (*above*) the flow engineers hope to accomplish with the CERP plan.

A 2015 photo of a stretch of the C-38 Canal (the Kissimmee River) that had not yet been altered by the restoration effort. Note the severe straight-line nature of the waterway.

This meandering portion of the Kissimmee River has been restored to its more natural flow and shape.

Above: The Orlando Wetlands Park.

Left: The Orlando Wetlands Park provides wildlife habitat.

Brian Lapointe, a research professor at Florida Atlantic University's Harbor Branch Oceanographic Institute in Ft. Pierce, has studied the sources of nutrient pollution in the Indian River Lagoon and elsewhere. Here on August 1, 2012, he poses at Sea Gardens, Green Turtle Cay, Abacos, Bahamas.

A tame great egret in Cedar Key, Florida, 2009.

The Kissimmee River in late afternoon.

8

Whose Water Is It?

IT'S EASY TO TAKE fresh water for granted. That's because it's so common. It seems to be everywhere. There are 330 million cubic miles of the stuff around the world.

Water, of course, is also utterly indispensable. It is the building block of the living cell. That means it's also the building block of the human body. In fact, it could be said water is the building block of all life on the earth and maybe of the universe, too.

Everyone needs water to survive. Just try to go without drinking it for three days. At that point you start to die. "Thousands have lived without love," wrote the poet W. H. Auden, "not one without water." Most schoolchildren know that the number-one ingredient of a human being is water. An infant is about 75 percent water. By adulthood that figure drops to about 60 percent. The three-pound human brain is 73 percent water.

So if clean, fresh water is so vital, do you have a *human right* to it?

The answer isn't clear. In February 2017 Pope Francis publicly declared, "All people have a right to safe drinking water." But that's just his opinion.

Article 25 of the Universal Declaration of Human Rights is a bit vague on the topic: "Everyone has the right to a standard of living adequate for the health and well-being . . ." It would be hard, of course, to be healthy without water.

In July 2010 the United Nations passed a resolution recognizing "the right to safe and clean drinking water and sanitation as a human right that is essential for the full enjoyment of life and all human rights." The General Assembly, however, soft-pedaled the resolution by also declaring that nations must take "immediate steps within their means towards the fulfillment of these rights." Thus, government should provide water to citizens—if it can.

A few international bodies use stronger language. The Council of Europe, for instance, insists that all humans have the right to enough water to satisfy their basic needs. At the 2007 Asia Summit participants asserted people do have the right to safe drinking water and basic sanitation.

A similar guarantee is hard to find in the United States. There's nothing about drinking water in the Bill of Rights. The Declaration of Independence asserts the right to "Life, Liberty and the Pursuit of Happiness," but it is mum about water. Of course, you could argue without water you couldn't have "Life."

In 2013 California became the only state to affirm that its citizens have the "right to safe, clean, affordable, and accessible water adequate for human consumption, cooking, and sanitary purposes." This doesn't mean, however, Californians are somehow entitled to get a case of bottled water from Sacramento, as if it were an armful of library books. Instead, the aim of the legislation is to encourage California's various government agencies to implement the idea of water as a right in their public policies.

Americans, however, may be slowly creating for themselves a right to water, even if no existing legislation says so. That's the opinion of Harvard legal expert Cass Sunstein. According to the professor, these "entitlements" for U.S. citizens emerge now and then from tradition and custom and become so widely cherished that no American president would dare violate them. Once widely accepted, they have almost the same firepower as a constitutional right.

Scholars call them constitutive commitments. In Sunstein's view they include, among other things, "Access to affordable water for drinking, hygiene, and sanitation."

Sharmila L. Murthy, a professor at Boston's Suffolk Law School, thinks an example of such a commitment was evident in the wake of the Flint, Michigan, lead-contaminated drinking water crisis. "Water holds a special, but overlooked, place in our culture, history, and laws," she writes, "and the widespread protests and outrage at the Detroit water shutoffs suggest that people perceive access to water as a right."

But perceptions never make anything legally enforceable. So far, Florida's legal code doesn't offer much about water as a human right or a commitment.

However, it has a lot to say about other water matters, which you might expect in a peninsula plastered with rivers, lakes, swamps, marshes, canals, and swimming pools. In Florida's early days water was everywhere, and the need for laws governing who gets to divvy it up would have seemed silly. In recent years, however, as people poured into the state, attitudes changed, and Floridians demanded new laws to control their behavior because they, like anyone else, weren't good at sharing resources.

The Tragedy of the Commons

This problem was immortalized by Garrett Hardin, the renowned ecologist and philosopher at the University of California. Hardin is best known for his 1968 essay "The Tragedy of the Commons." Hardin asks his readers to imagine a pasture that's free to anyone who herds cattle. What's in the best economic interest of each herder? Answer: To keep introducing one more cow after another to the pasture to eat as much as possible, as long as it's possible to do so.

The trouble is, of course, if all herders maximize their own self-interests, the pasture is destroyed. "Therein is the tragedy," Hardin writes. "Each [person] is locked into a system that compels him to increase his herd without limit—in a world that is limited. Ruin is the destination toward which all men rush, each pursuing his own best interest in a society that believes in the freedom of the commons. Freedom in a commons brings ruin to all."

Hardin's parable illustrates what's happening to many watersheds in Florida and around the world. Tracy Straub, chief engineer for the government of Marion County—one of the state's water-rich counties—sees the phenomenon unfolding where she lives and works. Because South and Central Florida are reaching the point of being built-out, she says, there is a drive to open up the northern part of the state to development, or, at least, to tap its water sources. As these North Florida counties develop, they sink new wells to get the water, which all comes from the same aquifer. Use the water or someone else will.

"It's a horrible position to be in if you're an elected official," Straub says. "A county commissioner who lives in North Florida might think, 'I have to protect my local water sources.'"

But you can't simply let this groundwater remain in place for future growth, she explains, because it is being pumped now by people in neighboring counties who are all pulling from the same aquifer. There's a good chance their usage means you won't have water when you need it.

As Florida, along with the rest of world, approaches the coming era of water limits, questions are emerging that are increasingly being decided by judges and attorneys versed in Florida water law. These questions include: Who owns the water in modern Florida? Should it be "free" like air? By whose rules do we decide how to divvy up the water pie?

Answers to such thorny questions lie in America's convoluted environmental movement history.

A Wild Ride

In the late 1960s, Florida, like the rest of America, was jarred by a raucous, anarchic period of social unrest, fueled by a militant antiwar movement and an aggressive civil rights crusade. Americans struggled over many vexing issues. One of the most pressing of these—and somewhat new as a topic of national debate—was what to do about an ailing environment. There were also three differing ideologies to choose from that dealt with such a question. Two had been around for a while, and the third, like a lot of notions appearing in the 1960s, was novel.

All had roots in the early 1900s. One was laissez-faire capitalism, the creed favored by huge monopolies and big business tycoons, which championed private property rights and free markets and scorned government regulation. Its "leave the economy alone" adherents argued that landowners, including mining and timber interests, should be able to do as they pleased with their property and the resources on them without interference.

Disagreeing with them were the conservationists. Their ranks included Americans from many different backgrounds, but what all had in common was opposition to the ongoing rampant and wanton industrial-scaled destruction of the nation's natural resources, unleashed by market forces. It also made economic sense, conservationists argued, to protect and renew your assets.

Thus, conservationism appealed to both nature lovers and business people. As policy and philosophy, it gained influence and stature throughout much of the first half of the twentieth century.

But by the 1960s, a more radical way of thinking about the natural world emerged: environmentalism. It, too, had roots in the early twentieth century, but they traced directly to the brain of Teddy Roosevelt's contemporary John Muir, a bearded naturalist, writer, and founder of the Sierra Club, today one of the most powerful environmental organizations in the country. Muir was also a chief crusader in the battle to persuade Congress to establish national preserves and the national park service. In his many books, articles, and speeches Muir demanded not just conservation but "preservation" of natural resources. The government should, he insisted, be not only the protector of areas already being exploited but the guardian of all places primeval.

In the unstable 1960s a new breed of activists tried to incorporate many of Muir's ideas into public policy. Many were also inspired by science writer Rachel Carson, whose book *Silent Spring* chronicled the toll modern chemicals were taking on the natural world. Her work shocked the nation and fueled America's burgeoning environmental movement.

The environmentalists used a new vocabulary, too, including words such as "ecology" and "interdependence of biological systems." If you want to keep nature intact, they warned, understand how nature functions and keep it whole!

By the early 1970s, some environmentalists morphed into *deep ecologists* who tethered themselves to a new and more radical ethic. Humanity had a greater duty, they argued, than just being good stewards of nature. All living things also had a right to exist, and this right didn't rely on their utilitarian or aesthetic importance to humans.

And water? It too was being examined through a new lens. No longer should it be considered as mostly a capital asset for business and agriculture. Nor, as was the case in Florida, should it be considered a nuisance to be drained away by canals and pipes at the expense of the natural world for the convenience of developers.

Complex hydrological functions were at play, insisted the activists. Water, wherever and whatever form it took in nature, was part of the same grand cycle and flow. And yes, all living things had a right to it, not just people.

More than just words were stirring things up. By this time a small-screen television sat in the living room of almost every American household. As never before, people across the nation saw powerful images on their living room screens of faraway happenings that often left a great impact on them. In 1969 few could ever forget seeing footage of the Cuyahoga River in Cleveland on fire. Actually, the oil and industrial waste that coated the river had burst into flames, not the water. Nonetheless, the image and the incomprehensible notion of a burning river swept across America, alarming and riling up millions.

By now, regardless of their philosophical leanings, most Americans could see clearly environmental decline across the national landscape. Floridians were no exception; bad news poured in almost daily about their home state: deadly algal blooms, sinkholes, salt water intrusion into drinking water, air saturated with carbon monoxide, acid rain, dying pelicans and bald eagles, and Florida panthers, bears, alligators, and crocodiles at the brink of extinction. There were horrific fish kills too. In 1969 Lake Thonotosassa in Hillsborough County, Florida, had made history for suffering the largest fish kill on record. An estimated 26.5 million fish died in a fiasco blamed on discharged wastes from nearby food-processing plants. Pensacola's Escambia Bay suffered repeated fish kills too.

But the problem was happening statewide and was so bad that on February 21, 1973, a *St. Petersburg Times* headline read: "Florida Wins Dubious Title of Nation's No. 1 Fish-Killer." The accompanying article revealed that, according to EPA data, "Florida now led the nation in fish killed by pollution in two of the last three years."

All this bad news along with a growing litany of other environmental horror stories prompted Americans everywhere to pressure their lawmakers to get serious about protecting the environment. Legislators responded with a raft of new laws designed to better protect the natural world. Among them were measures to safeguard endangered species and wild and scenic rivers. Others set aside funds to protect air quality, establish new national parks and wilderness areas, and regulate solid waste disposal. In 1965 Congress enacted the Federal Water Quality Act, which authorized the federal government to set water standards when state governments refused to do so.

Environmental angst even prompted Floridians to rework their state constitution. In 1968 they added Article II, Section 7, which spelled out the new commitment to the environment: "It shall be the policy of the state to conserve and protect its natural resources and scenic beauty. Adequate provision shall be made by law for the abatement of air and water pollution and of excessive and unnecessary noise and for the conservation and protection of natural resources."

Water as a Property Right

In 1971 Florida suffered a devastating drought. Freshwater wells in Miami were filling with salt water; fires burned everywhere; and the landscape turned brown and crisp. An alarmed Governor Reubin Askew summoned state leaders to an "American Assembly" in Miami and charged everyone in attendance to come up with new ways to protect Florida's waters. "It is time we stopped viewing our environment through prisms of profit, politics, geography, or local and personal pride," he told the gathering. "It is time for us to work together—to accept the truth about our problems in south Florida, and to set about solving them. It is time for us to do all of these things—because you know as well as I that the alternative will be disastrous to our economy as well as to our environment."

The state lawmakers rose to the challenge; the next year, they passed the Water Resources Act (WRA), a piece of legislation influenced by *A Model Water Code*, a white paper exploring the relationship between existing Florida law and water issues, recently crafted by Frank E. Maloney, a professor at the University of Florida law school, along with several of his colleagues. The code pertained not just to questions about consumptive use, but also other water issues, such as those dealing with navigation, dams, public access, and federal and state jurisdiction matters.

Up until 1972, water in Florida was viewed as a property right. Now that concept was headed for trouble. The new water act defined water not as a property right but rather as a "public resource benefiting the entire state."

Under the new legal framework, no single community, municipality, or county government could claim a local water resource exclusively for itself. However, there was also a legal requirement for any water-strapped community to exhaust all its local resources first before it went looking beyond its borders for relief.

The United States has long had two basic approaches to water law. In a western state—say, California—water allocation is, simply put, on a first-come, first-served basis. Water can be taken from a source as long as it is for a beneficial use.

Florida water allocation is founded on Eastern Water Law, which is rooted in the common-law riparian (the banks of a river) doctrine. Under this principle, property owners who share a freshwater lake or a riverbank have a right to extract a portion of the available water, providing their withdrawal doesn't hurt the common good. This right also extends to groundwater as well.

After the passage of the WRA, many Floridians were confused about whether they owned the water underneath their land in the same way they might own oil or minerals. After all, it seemed clear that once property owners pumped up groundwater from a private well, they could use it as they pleased. Didn't this also mean they owned the water when it was still in the ground?

The answer to this question was cleared up in 1979, when the Florida Supreme Court rendered its momentous *Tequesta* decision, which emerged from a dispute between a local community and a developer over access to water. The court decided that landowners can *use* the water underneath their land, but they don't *own* it until it has been extracted.

In a speech to a gathering of Florida water trade association professionals in 1982, Tampa attorney L. M. Blain explained it this way: "Who owns Florida's water? Whoever has it in his possession and under his control . . . and he owns it only as long as he has it in his possession and under his control. If his bucket or pipe leaks, he loses possession . . . and control . . . and no longer owns it."

The court's ruling also clarified that water was different from oil and minerals, because it flows across property lines, like air. Water is also a flowing, evaporating, condensing, seeping, oozing, freezing thing. You may not even own your own sweat. It vaporizes according to the laws of nature, not Florida.

A popular way to illustrate the government's new role after the *Tequesta* decision was to think of quail, deer, or gators. They and other wildlife wander across property lines all the time. Just because they temporarily exist on your property, however, doesn't mean you own them. That is, unless you shoot them. But to do that, you first must obtain the proper hunting license. The state, however, also has rules concerning how, when, where, and if such hunts can take place.

They also have created rules about using water—lots of them.

Rules of the Game

When state lawmakers created the water management districts in 1972—often called the Year of the Environment—they also enacted the Land Conservation Act, which allowed the state to sell state bonds to fund the acquisition of environmentally endangered lands.

They also passed the Environmental Land and Water Management Act and the Comprehensive Planning Act, both of which gave the state the authority to review and regulate major development projects that could cause a negative impact on regional resources.

In that same year, Congress signed off on the Clean Water Act. Its ambitious goal was nothing less than restoring the nation's waterways and recapturing their chemical, physical, and biological integrity. But the act required the states themselves to do the work. In Florida, today this responsibility falls to the DEP. First, the department has to identify polluted or impaired water bodies. Next, it estimates the daily amounts, or loads, of particular pollutants that these water bodies could take in each day and still meet water quality standards. That amount is called the total maximum daily load (TMDL).

To make sure there was consistency in how these assessments were carried out, state regulators were required to use scientifically established criteria. Those water bodies failing to meet the criteria were classified as impaired.

In 1974 Congress forged the Safe Drinking Water Act (SDWA) to protect the nation's drinking water supplies from pollution, whether it was caused by humans or nature. It was up to the EPA to set up health-based standards for all states to use.

Three years later, Floridians saw the arrival of the Florida Safe Drinking Water Act of 1977, which adopted enforceable statewide drinking water regulations.

In 1983 Florida lawmakers produced the Water Quality Assurance Act, which tackled the problem of hazardous waste leaking from storage tanks, abandoned wells, and septic tanks into potable water supplies.

This was followed in 1987 with approval of the Grizzle-Figg Act, which outlawed the practice of dumping raw sewage into "any river, stream, channel, canal, bay, bayou, sound, or other water tributary . . . without providing advanced waste treatment."

In that same year the Surface Water Improvement and Management (SWIM) Act appeared. It empowers the water management districts to restore water quality in polluted water bodies.

More laws kept materializing that dealt with water quality issues in the state, including the big restoration of the Everglades.

This ferment of legislative creativity gave Florida some of the most powerful means in the nation for imposing order on sprawl and dealing with water quantity and quality issues. To many activists it looked as if their home state—and maybe the rest of the nation as well—had at long last powerful tools to clean up the environmental mess.

Their euphoria turned to worry, however, when it became painfully clear that passing new laws was much easier than implementing and enforcing them. In Florida and across the country, people began questioning why there was a delay. Was the new legislation simply hard to carry out? Could it be, asserted some conservatives, that government bureaucracy was too inefficient? Or was it, as cynics insisted, that big-time polluters' allies in government were thwarting implementation of the new laws?

In the 1990s this frustration kept growing. Finally, responding to public pressure, more than thirty-five states and environmental groups took turns suing the EPA for not vigorously enforcing the law. Florida was one of the key courtroom battleground states. It was here in July 2008, in fact, that Earthjustice filed a lawsuit alleging the EPA and DEP were stonewalling, and the EPA had yet to develop clear, measurable, specific standards for water quality enforcement.

The challenges paid off. On January 14, 2010, EPA administrator Lisa Jackson signed a regulatory rule that set up "Water Quality Standards for the State of Florida's Lakes and Flowing Waters."

Finally, it seemed, real progress was being made. The EPA's new rules affected just about anybody or group that could be singled out for discharging pollution into a Florida water body, including industries, agricultural operations, municipalities, wastewater utilities, private companies, and other polluters.

Losing Ground

Until, that is, 2011. By this time the Great Recession was in full bloom, and Governor Rick Scott, along with a phalanx of pro-business figures, was firmly in power in Tallahassee, busily dismantling environmental protections and growth restrictions. They were also issuing new rules, but these were not eco-friendly. No longer, for instance, were developers required to show their projects would cause no harm to the environment. Instead, opponents had to show the developments would cause harm.

This one change made it easier for developers to obtain water withdrawal permits. The state legislature also gave itself the authority to restrict water management district budgets. According to a special report prepared by 1000 Friends of Florida, the level of revenue for Florida's water districts dropped 39.4 percent between 2011 and 2012, leaving them with $703.3 million less to protect water sources and prevent flooding. Such interference was the very thing Governor Askew had feared.

The water districts also suffered staff reductions of 596 people and slashed benefits. According to published reports and personal interviews, many of these firings and demotions of career staffers were carried out to remove individuals seen as obstacles to new development. Many former and current state employees said these purges came swiftly and uncaringly. "What hurt the most," recalls one high-ranking career DEP administrator, "is that many of those fired were notified by e-mail."

"How," asked the authors of a special 1000 Friends report, "will water management districts, in a fast-growing state with many existing water problems, be able to retain their capabilities to properly manage and protect water supplies, provide flood protection, and manage and protect natural resources with severely diminished funding and major staff reductions? The next major hurricane or drought will be another reminder of lessons learned from the past that resulted in the 1972 water management system."

State Senator Linda Stewart, a Democrat from Orlando, also laments the loss of talent of scientists and engineers in those purges. "The pool of knowledge in the districts and the DEP has been diminished," she said. "People who had years of experience working with the environment and developers is also diminished. They knew too much. They were replaced with too many 'yes' people who make the decisions—and that's why we're going down the wrong road."

Supporters of the new regime's actions, however, argued they were much-needed reforms. Lowering taxes and reducing obstacles to growth and other meddlesome regulations, they claimed, would create jobs and put Floridians on the road to prosperity.

Political and legal wrangling over such issues, of course, is embedded in American culture. Even in the golden years when Florida had pro-environment governors and functioning regulatory agencies, there were still big and costly and divisive legal conflicts over the environmental issues.

Many of those fights were spats over water, but a few are ongoing, fully developed Florida water wars. More conflicts may be on the way.

9

Florida's Water Wars

OIL WAS THE SOURCE of much conflict and war in the twentieth century. Will people also fight over water in the twenty-first?

Many international experts predict that's exactly what's going to happen. "We judge that the use of water as a weapon will become more common during the next 10 years with more powerful upstream nations impeding or cutting off downstream flow," write the authors of a 2012 National Intelligence Council report on global water security. "Water will also be used within states to pressure populations and suppress separatist elements." They also predict the growing possibility of terrorists' targeting their enemies' dams.

The report anticipates special foreign policy challenges for the U.S.: "During the next 10 years, many countries important to the United States will experience water problems—shortages, poor water quality, or floods—that will risk instability and state failure, increase regional tensions, and distract them from working with the United States on important US policy objectives." A few likely hotspots are the water basins of the Nile, Tigris-Euphrates, Mekong, Jordan, Indus, Brahmaputra, and Amu Darya.

Agriculture is already the world's biggest user of water and fertilizers, and it's still hard to feed everyone on the planet. A growing population worldwide will put even greater pressure on the world's farmers to provide extra food. And if future populations insist on eating the same high-animal-protein diet that most Americans now enjoy, a lot more cattle will be required. This in turn is likely to boost demand for water to irrigate crops that animals eat, clearly an unsustainable thing to do, when you consider beef production requires about ten times more water than what's needed to produce the same amount of calories from cereal and vegetables.

Rising demand for electricity to satisfy the multiplicity of modern needs of a growing world population and ever-urbanizing world will also tax water reserves.

The National Intelligence Council report also warns that if humans deplete the world's rivers to satisfy these new power demands, the world community's ability to produce hydroelectricity will decline, possibly disrupting the global economy.

Water problems alone aren't expected to cause any one state to collapse, but water problems—along with other debilitating troubles such as poverty, overcrowded living conditions, malnourishment, and poor public health—could push a distressed state past a tipping point.

Water-rich Florida would seem an unlikely place to see any of these dark prophesies unfold. However, for almost a half century, Floridians have wrangled over water issues. They have fought—and still do—over access to water, water pollution, the issuing of water consumption permits, drying-up springs, and scores of other water-related matters. In fact, local governments, utilities, activists, and developers challenge one another all the time.

The stakes are often high in these encounters, as the state's first major water battle demonstrated. At the time, it shook Florida and the nation.

Jettisoning the Jetport

In the 1960s the Florida environmental movement was in its infancy, when a juggernaut of political and commercial interests made clear its plans to build a major jetport in the Big Cypress Swamp, a region of huge wetlands in Southwest Florida about the size of Delaware. Florida's development community was ecstatic, but a growing number of Floridians feared that if this project succeeded, both the Big Cypress Swamp and the Everglades were doomed—and so, too, the recharge supply of South Florida's drinking water.

Although the Big Cypress Swamp neighbors the Everglades, it has a different bio-scape. For one thing, Big Cypress's turf is a foot to two feet higher than the Everglades, which allows more hardwood trees to grow. For years, conservationists tried in vain to include the entire Big Cypress Swamp in the Everglades National Park. Instead, it became a place to exploit: loggers chopped down centuries-old cypress trees, while developers cleared land and blazed dirt roads through the wild growth.

The jetport master plan envisioned building a futuristic aircraft facility—the biggest in the world—capable of handling jumbo jets, each capable of carrying one thousand passengers on a single flight and travel at supersonic speeds. Planners also wanted a high-speed monorail and a new interstate highway system that sliced through wetlands and connected the jetport with Florida's population centers. If all went as planned, there'd be a new city of 150,000 people, planted right there in Florida's interior watery wilderness.

Why build it in the Big Cypress Swamp anyhow? The answer came from the Dade County Port Authority, which had concluded nobody in Florida's growing urban areas was going to put up with the sound barrier being broken all day long by supersonic jets over their own neighborhoods. So officials chose a remote thirty-nine-square-mile tract in the Big Cypress instead, just six miles away from the Everglades National Park. Who would complain, except a few hundred Native Americans who lived nearby?

Plans were made. Permits approved. Work got under way in 1968. Developers and investors were confident it was just a matter of time until the modern world blasted into the Big Cypress Swamp.

But there was a snag. By now winds of change were blowing across America. Standing up for nature had become a new and trendy pastime for many Floridians, a call of duty for the sincere.

Opponents of the jetport mobilized all over the state. Meanwhile, many employees within the federal government were experiencing doubts about the wisdom of building a modern airport in the heart of one of the world's most biologically diverse areas and Florida's water supply. At the behest of the U.S. Interior Department, a trio of scientists—Luna B. Leopold, a senior research hydrologist (and son of author Aldo Leopold); Arthur Marshall, a biologist and combat officer who experienced D-Day; and Manuel Morris of the National Park

Service—produced a study of the potential environmental impact of the jetport. It may have been the first ever such study done in Florida, or anywhere in the United States.

Their work debuted on September 17, 1969, and provided a scathing rebuke of the project: "Development of the proposed jetport and its attendant facilities will lead to land drainage and development for agriculture, transportation, and services in the Big Cypress Swamp which will inexorably destroy the south Florida ecosystem and thus the Everglades National Park."

The growing anti-jetport movement—an unlikely coalition of hunters, politicians, Native Americans, and conservationists—now had the science it needed to be legitimate.

At first, jetport supporters weren't overly concerned about the opposition. In fact, some seemed amused and annoyed by those they dismissed as silly nature lovers standing in the path of progress. "Alligators make nice shoes and pocketbooks," was the wise-guy retort that Michael O'Neil, Florida's Department of Transportation secretary, gave when asked by journalists if he was worried about environmental damage the jetport posed to the Everglades National Park. "I'm not really concerned about the alligators," he added. "And I don't miss seeing the dinosaur either."

Another official haughtily dismissed jetport critics as nothing more than butterfly chasers. Richard H. Judy, a high ranker at the Dade County Port Authority, took a different approach, and piously intoned that the transportation project was in keeping with America's duty to "meet our responsibilities and the responsibilities of all men to exercise dominion over the land, sea, and air above us as the higher order of man intends."

In Tallahassee, the corridors of power had long echoed with the sounds of jetport endorsement. However, there were also a few individuals experiencing second thoughts after having read the impact study. Nathaniel Reed, an aide to Governor Claude Kirk Jr., was one of them. After sharing his concern about probable massive environmental devastation with his boss, Reed later stood before Congress, urging a halt to the project. Kirk did too.

But President Richard Nixon is the man who killed the project. After reviewing a U.S. Interior Department study opposed to the project,

Nixon ordered an end to jetport construction. Instead, on February 8, 1972, in a special message to Congress, he proposed legislation to create the Big Cypress National Fresh Water Reserve.

That goal came into focus in 1973, when Kirk's gubernatorial successor, Reubin Askew, led a successful campaign in Tallahassee to convince the state legislature to pony up $40 million to help buy Big Cypress and protect it. A year later, Congress came up with an extra $150 million to help defray the cost. In October of that year, with the help of President Gerald Ford, the 729,000-acre vast freshwater swamp known as Big Cypress finally became a federally protected national preserve.

Saving the Big Cypress was a spectacular win. But as all parties in the feud decompressed, another Florida water war was heating up in the Tampa Bay area.

Intra-County Warfare

It was on the Gulf Coast where the cities of Tampa, St. Petersburg, Clearwater, and other smaller communities converge that one of Florida's biggest and most rancorous conflicts over water came to life in the 1970s.

Salt water had been creeping into the well fields of fast-developing Pinellas County for decades, which made it increasingly harder for water utilities to provide water for newcomers. Pinellas officials decided to solve the problem by increasing extractions from the well fields their county owned in the nearby counties of Hillsborough and Pasco.

After a while, wetlands and lakes in those two counties began disappearing. There was also an increase in the number of sinkholes—those yawning surface cave-ins that occur suddenly when the limestone rock underneath collapses. Sinkholes are dangerous, as anyone who's lived in the state long enough knows; they've swallowed cars, dredging machines, houses, and even people.

To the residents of Hillsborough and Pasco Counties, these occurrences suggested just one thing: underground water tables were sinking. And they knew who to blame.

Not everyone, however, was willing to admit Pinellas County's water pumping was at fault. In fact, some of that county's officials instead blamed a current drought. Just wait until it rains again, they argued; then the water table would rise.

That argument did little to appease the increasingly grumpy grievants. Nor did it win over Swiftmud officials, who ordered Pinellas's utilities to cut back on groundwater extraction and find other water supplies.

This decision was a blockbuster. It sent Pinellas's builders, realtors, and political figures scurrying into crisis mode. Curtailing water supplies, they complained, would slow growth and hurt the local economy! How could they do this to us? Pinellas and St. Petersburg officials argued they'd already spent a lot of money on infrastructure needed to transfer water to their users. Not only that, they owned permits to pump the water. A deal was a deal!

Swiftmud, however, didn't back down. That meant battle lines were drawn, lawyers recruited. The coming years saw a bitter mash-up of lawsuits and strained relations among local governments, water utilities, citizen groups, and the business community in the Tampa Bay area. Millions of dollars were spent on bickering in court and accomplished little.

Finally, in the 1990s the state government muscled its way into the fray and convinced all parties to forge a peace treaty. At last, a deal was struck that required the existing Tampa water authority be restructured and repurposed. It was given both expanded powers and a new name—Tampa Bay Water. Now it was a designated *regional* authority that could set water policy for a community of governments, including those of the cities of Tampa, St. Petersburg, and New Port Richey and the counties of Pinellas, Hillsborough, and Pasco. All factions pledged to work together to reduce groundwater pumping in the well fields north of the urban area and to develop new water supplies without hurting the environment.

Hopes ran high that Tampa Bay Water could supply enough water for everyone when its desalination facility and supply reservoir were up and running.

Since then, the new authority has kept the peace. But by early 2018 cracks in the agreement appeared when Tampa city officials began promoting the Tampa Augmentation Project, or TAP. City officials said they needed to develop new water sources on their own to meet future demands and be ready for the next drought. The project, if ever implemented into law, would let the city stop sending treated wastewater from a treatment plant into Hillsborough Bay. Instead, it would be piped to wetlands and rapid infiltration basins where it would be naturally cleansed before finding its way back to the Tampa Bay Bypass Canal, a water source for the city. An alternative version of the plan calls for storing water in an aquifer recharge and recovery well until it's needed.

Tampa Bay Water officials, however, opposed TAP, arguing it could unravel the existing agreement among the area cities and counties. Some skeptics worried that the plan didn't require enough safeguards to mop up pharmaceuticals and other contaminants.

In the spring of 2018, the legislature approved a bill that legitimized the kind of potable reuse projects Tampa and several other Florida cities had in mind, but the governor vetoed it. Critics of TAP were pleased, but supporters were left wondering where to find new water sources to keep up with expected growth.

Council of 100 Heats Things Up

There have been others who also have doubts about Tampa Bay Water's ability to provide enough water. For example, consider the Tampa-based Florida Council of 100, a statewide business group accustomed to giving Florida governors advice. In 2003 the private lobbying group, whose membership included CEOs and leaders from Florida's development, agriculture, and business sectors, commissioned its own water supply study and confidently handed it off to Governor Jeb Bush.

The truth about water, according to the council's findings, was that Florida had plenty of it. Just fly over the state and see for yourself how much there was down there.

No, what ailed Florida, insisted the report's authors, was a "water storage and distribution" problem; the sad and unfair truth was that the water simply wasn't distributed evenly across the state. As the study put it: "80 percent of the population and public consumption is south of I-4; 80 percent of the water resources are north of I-4."

Among the council's recommendations were what you'd expect any water commission to suggest: develop new water sources, use desalination, build new reservoirs, reclaim stormwater, and so forth. But the council also put forward another idea that proved eye-bulgingly controversial: Take water from those who have it, and give it to those who don't.

The council had another big idea that may have discomfited staffers at all the water management districts: "A statewide water distribution system would establish an economic value to water and water would become a general revenue source for the state of Florida and sending areas."

The authors added: "Many argue that a statewide water distribution system from water-rich areas to water-poor areas is more environmentally sound and cost effective than other alternative water supplies, such as desalination."

Questions popped up across the state like mushrooms overnight. Did the council say a statewide distribution center? A centralized waterczardom with the authority to pipe water from water-rich North Florida to Central and South Florida?

The idea of piping water from here to there wasn't outlandish. In fact, people have been doing it since the days of the Roman Empire. In recent years, a slew of pipeline projects has popped up across the United States, including schemes to pipe water to California and Arizona from as far away as Washington, Oregon, Alaska, and British Columbia. There have also been serious suggestions to tap the Great Lakes and pipe the water out west.

In Florida, in fact, an above-ground water pipe along U.S. 1 has been delivering water for decades from a well field at Florida City on the mainland to as far south as Key West, 130 miles away.

Still, what sounded so logical and fair to the council seemed like one huge, lousy idea to a great number of people in North Florida, many of whom lived in small towns and rural areas that were still sufficiently southern and cantankerous. Unbridled comments, such as that uttered by Clearwater realtor Lee Arnold, the head of the council study group, who dubbed the Suwanee River region as "the Saudi Arabia of Water," only made matters worse.

Across North Florida opposition rose quickly and militantly against any and all perceived water pirates down yonder. Many residents declared it was their water, not South Florida's! Giving into water bullies would only encourage more irresponsible development!

There were expressed fears of North Florida springs being depleted, sinkholes opening up. One Levy County official brazenly spoke of using arms to protect his county's local water. Others spoke darkly and openly about sabotaging water pipes, if Tampa foolishly dared to install any.

Faced with all this stinging criticism, the council didn't feign neutrality in the matter. Its chairman, Al Hoffman, one of the state's biggest developers, said, "We're not going to apologize for the fact that we are CEOs of the largest companies in Florida, that we are vital and involved. What do you think we are doing this for?"

When the ruckus rattled about much too long and loudly, state officials dispatched a Senate Natural Resources Committee to six North Florida community sites to hear what locals themselves had to say about the Florida Council of 100's water distribution scheme. It couldn't have been much fun for the committee members to listen to a long parade of residents vent their anger.

The last of these sessions was a night meeting on November 20, 2005, at a high school in the town of Chiefland, in rural Levy County. An estimated one thousand scowling men, women, and children arrived in cars, trucks, and buses from Levy, Alachua, Dixie, Marion, Gilchrist, and other counties. Their handmade signs expressed rising fury. "Over My Dead Body!" and "Our Water Is Not for Sale." One by one, protestors took turns blasting both the Council of 100 and its plan.

It didn't take long for Florida's power barons to realize that if one thousand angry people took the time to travel to a small town on a school night to rabble-rouse, surely many more kindred souls weren't too far behind.

In the following weeks, dozens of county commissions and city councils across North Florida voted to oppose the water redistribution plan. There was even discontent heard in Central Florida from residents who accused their own local political leaders of resorting to a water grab to mask their failure to manage local growth.

North Florida's protestors may have believed they were justifiably protecting their home turf, but they were wrong about the water being theirs. Water, says Florida law, is a state, not local, resource. And state law does provide for water transfer, but only as a last resort.

Still, their spontaneous protests had clearly communicated to everyone the strong value Floridians placed on the traditional roles water played in their lives.

When it became clear that the plan was a political loser, Miami developer-turned-governor Jeb Bush shelved it. Since then, instead, Florida has poured millions of dollars into water management districts to develop new water supplies.

But pipeline ideas never go away for long. In 2009 many of the Council of 100 ideas made a comeback, this time packaged in a state Senate Environmental Preservation and Conservation Committee interim report. One of them was eerily all too familiar: "Let us . . . establish a central regulatory commission that oversees Florida's water resources and supply development."

The report, however, never inspired any legislation. For now, residents of North Florida have relaxed their trigger fingers.

Where the Water Is

The Council of 100 may have lost its water grab, but greater Orlando's business leaders hope to do better. Ever since the news came out that their neck of the woods had just about overpumped local groundwater, they've sought new water supplies and said so, most recently in the Central Florida Water Initiative. At the moment, plans are being fine-tuned to drain millions of gallons a day from the surface waters of the St. Johns River to quench greater Orlando's growing urban thirst.

Several environmental groups are clearly upset about Orlando's siphoning plans and have worked up a long list of water-related problems they say could occur, if anybody would read them.

Orlando itself, however, faces a formidable foe to the north, perched downstream on the banks of the St. Johns River—the City of Jacksonville. The last thing the residents of this municipality, named after the seventh U.S. president, want is further degradation and depletion of the scenic river they depend on, which flows right through their town center.

And besides, Jacksonville has its own water pie to expand for its growing population, and the St. Johns is just so convenient.

Jacksonville, however, isn't above reproach when it comes to water consumption in the eyes of residents in neighboring counties who for years have complained that their groundwater tables keep dropping, as the pumps of the "World's Biggest City" keep sucking more and more of the region's groundwater.

Trouble in Palatka

As tensions rise among the municipalities, another North Florida water feud, which has been under way for a half century, shows no sign of disappearing. This battlefield is located about eighty miles southwest of Jacksonville on the Ocklawaha River, a tributary of the St. Johns. Here, in leafy Putnam County, sits the Rodman Dam and Reservoir, built decades ago for navigational purposes for the now-defunct Cross-Florida Barge Canal project. Although the long-forgotten canal will never be completed, the 9,500-acre pool created when the dam was built in the 1970s remains. Beneath the surface water of this flooded portion of the Ocklawaha lurks a murky, sunken graveyard of dead standing trees and stumps.

At first glance it is hard to understand why the four-gate spillway dam (renamed the George Kirkpatrick Dam to honor the Florida state senator who led an effort to protect the dam from obliteration) hasn't been removed. A big reason is that over the years the reservoir has evolved into a popular bass-fishing site. Supporters claim this outdoor pastime, along with other recreational opportunities, contributes to the local economy of a very poor county. The nonprofit group Save Rodman Reservoir, Inc., also thinks a new ecosystem that's worth saving has evolved over the years. "The barge canal may have been an ill-conceived project in the beginning," it concedes on its official webpage, "but just look at the incredible animal habitat and recreational resources we have today."

Here and there across North Florida, political figures pop up now and then who argue the pool should remain intact so that it could be used as an alternative water supply, presumably to grow somebody's water pie. That's not likely, say many water experts, because the reservoir water is far too polluted and unreliable.

For decades, reservoir preservationists have fought with opponents who want the facility breached so that the Ocklawaha River can run free and regain its former natural flow. Once that happens, argues the anti-reservoir crowd, migratory fish now denied free and open access can once again make their way from the St. Johns River to Silver Springs, as they had in the past. Anywhere from two hundred to three hundred manatees would also show up, predicts wetlands scientist Robin Lewis. Currently, only a few of the gentle, plodding mammals now make it through Rodman's gates.

Many of the river restorers are environmentalists, including spiritual heirs of Marjorie Carr, the Micanopy woman who spearheaded the campaign to stop the Cross-Florida Barge Canal. To them, the Rodman facility is an annoying reminder of unfinished business.

Money Matters

The fuss over the Rodman, however, isn't simply a struggle between river lovers and bass fisherman. Economics, as always, is also involved.

A central point is that Putnam County is poor. In 2015 the county was Florida's poorest, with a poverty rate of 26.4 percent, according to a *Wall Street Journal* review of U.S. census data. That same year the Robert Wood Johnson Foundation and the University of Wisconsin Population Health Institute published a study revealing "Putnam County has the worst health factors of the sixty-seven Florida counties."

Headlines like these don't spur the kind of economic growth that a poor county needs. Small wonder that many Putnam County residents don't want to let go of the reservoir that they believe is a source of revenue, as paltry as it may be.

Upstream, in Marion County, the economic situation isn't so dire. But the county's pride and joy, Silver Springs, is now a state park and no longer the economic engine it once was as the state's number-one tourist attraction with all its giraffes, monkey shows, and concerts.

Outside the park's entrance, the community of Silver Springs isn't doing very well either. In fact, in 2018, it became a Community Redevelopment Area. This means it was targeted as an economically distressed area needing extra county taxes for economic development.

However, many local Ocalans believe both the health of the springs and the local economy would get an extra boost if the Rodman is breached. As Robin Lewis suggests, more fish and manatees would mean more paying visitors to the local economy.

A restored Ocklawaha could prove beneficial to Putnam County, too. A 2017 University of Florida study concluded that the economic impact from ecotourists who use the natural portions of the Ocklawaha River in kayaks and canoes is twice that of the anglers and boaters at Rodman Reservoir.

For the moment, the issue festers in an ongoing stalemate. As both sides wait for something to happen, the facility continues to age and deteriorate. One day in the near future, state lawmakers—many of whom are likely to be Republicans—will have to decide whether to spend millions of taxpayer dollars to fix and maintain an aging relic of the New Deal, the legacy of Franklin D. Roosevelt, a Democratic Party superstar.

Three-State Water Wrangle

By 2017 another costlier and bigger North Florida water war was being slugged out in the U.S. Supreme Court with Florida and Alabama on one side and Georgia on the other.

All three states share a water system that starts out as two distinct rivers in Georgia and ultimately drains 19,500 square miles of land. The 344-mile-long Flint River is one of them. It seeps up from the earth in the West Central Georgia city of East Point, situated south of the sprawling Atlanta metro area. The river then flows underneath the runways of Hartsfield-Jackson Atlanta International Airport before heading south to rural West Georgia.

The Chattahoochee River, meanwhile, forms in Jacks Gap in North Georgia's sparsely populated Union County in the Blue Ridge Mountains. The river's 430-mile journey passes through Lake Lanier—a 38,000-acre artificial lake created in 1956, when the Army Corps of Engineers constructed Buford Dam, which lies in the river's headwaters north of Atlanta. Heading south, the Chattahoochee rushes through the sprawling Atlanta metro area, which extracts great quantities of water for a variety of municipal uses. But the river also receives inputs of urban water—much of it reclaimed. In fact, according to the Chattahoochee Riverkeeper, more than one hundred private and public utilities discharge in excess of 250 million gallons of treated wastewater every day back into the Hooch, as Georgians like to call their river. More is produced all the time. Coping with this increasing volume puts great stress on the Chattahoochee, whose watershed lies north of Atlanta and is also the smallest serving any metro area in the nation.

After the Hooch departs Atlanta, it veers far to the south and for a while forms the border between Georgia and Alabama. As it bends back to the east and flows through the agricultural belt of South Georgia, it loses millions of gallons to farmers who use the water to irrigate thousands of acres of peanuts, cotton, soybeans, corn, and other crops.

Near the Georgia and Florida state line, the Chattahoochee converges with the Flint at Lake Seminole, upstream from the Jim Woodruff Dam. From here on, this confluence forms the Apalachicola River—Florida's biggest river in terms of volume—and moves 106 miles southward through a floodplain before discharging into the Gulf of Mexico at Apalachicola Bay, historically one of the most productive in America.

For more than a century, residents of Apalachicola, a picturesque fishing village perched on the banks of the bay, have survived by harvesting the local estuary. It is home to Florida's prized oyster industry.

Over the decades, Apalachicola has endured hurricanes and periods of drought. But by the 1970s the community faced a new problem that seemed intractable. Both the Apalachicola River and the bay into which the river flows were showing signs of distress. Less water was now flowing, which meant the river's floodplain was drying out. Fewer nutrients were swept up by moving water and deposited into the estuary. A drop-off in fresh water volume also caused the salinity level to rise, which was bad news for oysters, other shellfish, and fish.

The river's decline also triggered a deterioration in relations between Florida and Georgia that grew worse in the coming decades. State officials in downstream Alabama teamed up with Florida and accused the Peach State of extracting too much water to support its own population growth and agriculture needs, thus depriving its downstream neighbors of their fair share of the water.

Georgia officials responded that Atlanta, like any city, has riparian rights because the river flows nearby. Don't forget, they say, that 80 percent of the river basin lies in Georgia, 14 percent in Florida, and 6 percent in Alabama. Therefore, it's only fair that Georgia should receive the lion's share of water. Any damage to Florida's oyster industry, insist the Georgians, was Florida's fault, not theirs. Besides, isn't it hypocritical to criticize Georgia while Florida's political class is encouraging the Florida growth machine to suck up more and more groundwater?

For decades, such reasoning appeared in courts and news accounts, along with ample doses of political grandstanding and finger-pointing.

Headline writers like to feature the struggle as a modern tri-state water war. But they misstate the issue, says Steve Leitman, a Tallahassee-based environmental hydrologist who has studied the Apalachicola River for decades. "Apalachicola isn't really a war between three states," he says. "It's a war between a suite of different interests in three states, such as mining, environment, navigation, farming, flood control, and urban growth."

Nor is this a David and Goliath story—that is, a tiny Florida fishing community up against the huge Atlanta metropolis, explains Dan Tonsmeire, the Apalachicola Riverkeeper. "There's an agriculture component to this conflict that's larger than Atlanta's consumption. Georgia farmers use more water than Atlanta does in one day. There are also five federal reservoirs on the Chattahoochee, and they evaporate as much water as Atlanta uses each day."

The amount of flow isn't the only environmental issue. There's also concern about the timing of the water flow, an essential component of river ecology. Upstream reductions of flow during times of drought or for other purposes could disrupt the hydrology and ecology of the Apalachicola River.

This multistate dispute was clearly an interstate problem, thus enabling the federal government to dive in. In 1997 Congress set up the Apalachicola-Chattahoochee-Flint Basin, known as the ACF Compact, which required the three states to work to negotiate a solution by a set deadline.

Nothing worked.

A New Player

In 2000 a new player entered the fray. It was Southeastern Federal Power Customers, Inc. (SeFPC), a group of electric cooperatives and municipal power companies that sued the Corps, asserting that Atlanta wasn't supposed to be receiving water from Lake Lanier in the first place. Congress authorized the dam to be built, the group alleged, for the purposes of flood control, navigation, and hydroelectric power, not as a water supply for Atlanta.

But a water supply is exactly what the Corps had allowed. Today, nearly four million Georgians, including about 70 percent of the residents of metropolitan Atlanta, dip into the Hooch for their personal, industrial, and business uses.

In 2009 things seemed to be headed Florida's way when a federal district court agreed the Corps lacked authority to use the lake as a water supply. The court also instructed the three disputing states to figure out a mutually agreeable water-sharing plan and obtain approval from Congress.

Three years later, however, the U.S. Court of Appeals in the 11th Circuit reversed the earlier decision, ruling the Corps did possess the authority. It also told the Corps that it had one year to figure out how much water it could send to Atlanta and still meet its other responsibilities. This decision was one more blow to tiny Apalachicola, which produces 90 percent of Florida's oyster crop and 10 percent of the nation's; it is also nearing the brink of collapse.

In 2013 Georgia asked the Corps for even more water, pushing up its desired daily withdrawal from the river system to 593 million gallons.

In August of that year, Governor Rick Scott announced the State of Florida was taking legal action against Georgia to check its water consumption. This action allowed Scott to make the tri-state water dispute a cause célèbre. In a public statement, he pointed out the metro Atlanta area was withdrawing 360 million gallons per day from the Chattahoochee River. "Meanwhile, Georgia's consumption is expected to nearly double to 705 million gallons per day by 2035, as Atlanta's population and water consumption grow unchecked," the governor said. "That estimated daily consumption represents the approximate water volume of the entire Apalachicola Bay."

The State of Alabama filed an amicus brief in support of Florida. Alabama has an understandable self-interest in the matter. State officials fear many vital aspects of life in their state would be jeopardized if Georgia's water consumption isn't curtailed. The ACF Basin provides sustenance to municipal water supplies, fisheries, and wildlife habitat. Its recreational sites need that water, and so do many Alabama electric power generation plants, industries, and farms. Navigation also could be disrupted if Alabama's fair share of the ACF is reduced.

Florida's Bad Day in Court

Florida officials were hopeful, but on February 14, 2017, they received a bitter Valentine from Ralph Lancaster, the special master appointed by the U.S. Supreme Court in the wake of Florida's 2013 lawsuit.

Florida had lost. Lancaster recommended to the U.S. Supreme Court, which would have the final say, to deny Florida's request because the state hadn't proven that capping Georgia's use of water would help alleviate water problems in Florida.

The judicial official, however, also wondered why Florida hadn't included the Corps of Engineers in its suit. "[W]ithout the Corps as a party, the Court cannot order the Corps to take any particular action that helps Florida," he observed.

Lancaster also acknowledged that the Apalachicola River was in distress and that Georgia was acting irresponsibly. "There is little question that Florida has suffered harm from decreased flows in the River. Florida experienced an unprecedented collapse of its oyster fisheries in 2012. It also appears that Georgia's upstream agricultural water use has been—and continues to be—largely unrestrained."

That wasn't all. The special master further wrote, "Georgia's position—practically, politically, and legally—can be summarized as follows: Georgia's agricultural water use should be subject to no limitations, regardless of the long-term consequences for the Basin."

Despite this blistering rebuke, Georgia state officials were ecstatic. So, too, were the Atlanta business community and South Georgia farmers. After all, they'd won, at least until the Supreme Court spoke.

In Tallahassee there was disappointment. Some lawmakers were also grumbling about the price tag on the state's unsuccessful legal action. According to a *Miami Herald* and *Tampa Bay Times* analysis, the legal cost was $98 million.

So, for the time being, Georgia seemed the victor. However, in June 2018 the U.S. Supreme Court overturned Lawrence's recommendation and sent the case back to him. Florida was to get another chance to prove its case.

This time, Florida had a win! But had it really? No matter who won in court, the river basin was the real loser.

"Today we're not getting it right," rues Riverkeeper Tonsmeire, a man who lists on his résumé stints as an Idaho river guide, Alaska fisherman, and Nature Conservancy employee. This water veteran views the relentless assault on the iconic river as an "indicator" of a bigger threat.

Once people ignore that indicator, Tonsmeire says, they take the river for granted; one day, it will be gone, putting at risk yet one more slice of the natural world, which is "the basis of human existence."

Despite all the setbacks, Tonsmeire tries to keep his chin up. "There is a path for this problem and others," he insists, "and it's to stop exploiting nature. Treat water as something precious, not just a commodity."

As residents of three southern states waited for yet another court battle, a new water squabble was under way in Central Florida. There, a regional authority that provides drinking water to the counties of Sarasota, Charlotte, and DeSoto and the city of North Port was seeking a fifty-year permit to increase withdrawals from the Peace River up to 258 million gallons daily. However, a regional water cooperative in upstream Polk County—worried about that county's own future water needs—was headed to court to halt the permit from being issued.

Meanwhile, another, and much older, Florida water war was heating up—as it had repeatedly over the decades. This one had been festering in the heart of the Everglades. It is the mother of all Florida water wars.

10

The Mother of All Florida Water Wars

FLORIDIANS HAVE KNOWN for a long time that they've been slowly killing the Everglades. In 1947 Miami journalist, author, and environmentalist Marjory Stoneman Douglas revealed in her now-classic book *The Everglades: River of Grass* that this huge sheet of water wasn't a swampy waste like many people once thought; instead, it is a vital river in need of urgent protection and rehabilitation from the efforts to drain much of it to sea.

In 1976 the world-famous swamp received renewed public awareness when the Everglades National Park became an International Biosphere Reserve. Three years later it was christened an International World Heritage Site. Such accolades were more than praise. They legitimized what many environmentalists, scientists, and politicians had been saying about the fragility and ecological importance of the Everglades.

The jetport battle of the 1970s had also awakened much of the American public to the reality that the Everglades and its national park were in big trouble. By this time the big swamp's problems were so dire that many of its defenders claimed it needed to be put on life support—and they weren't exaggerating.

For one thing, the Everglades was now half its original size of eighteen thousand square miles. Wading bird populations had been reduced between 90 to 95 percent, and sixty-eight plants and animals were now threatened or endangered.

The mighty river's sheet flow had been fragmented and siphoned off in a myriad of directions to farms, urban areas, and the sea. It was also being blocked by the Everglades Agricultural Area, the Tamiami Trail, and South Florida's westward-creeping urban sprawl.

Florida's burgeoning human population had also introduced many exotic life forms into the state. Nonnative trees, for instance, such as Australian pines, melaleuca, and Brazilian pepper trees were racing across the Everglades, damaging wetlands and inflicting other environmental disruptions. South Florida's canals teemed with exotic fish such as clown knife fish, Mayan cichlids, and walking catfish. There were scary new predators, too, such as Burmese pythons and Nile monitor lizards roaming the swamps.

The Everglades' wetland muck soil was in trouble, too. Under normal conditions this organic stuff spends a lot of time underwater. Once water is drained, however, the muck dries out and shrinks in air and sunlight. It also oxidizes. Add to that the forces of wind erosion, burning, farming, and a loss of groundwater, and the soil starts to subside, which means it travels downward. Some scientists think that as much as six feet of soil in the Everglades farmlands have subsided since the drainage glory days, with oxidation responsible for half of the loss. It may be the case that subsidence will keep going until it hits bedrock.

Everglades soil is also low in phosphorus. That has never been a problem for the natural flora and fauna that had adapted over thousands of years to the natural conditions of the Great Swamp.

Florida sugar cane, which is grown in abundance in the Everglades, though, needs phosphorus. That's why the area's sugar farmers feed their crops phosphorus-based fertilizer. The problem is that excess amounts of this nutrient also spread into the surrounding area. "Under natural conditions water flowing into the Everglades would contain 8–10 parts per billion (ppb) of phosphorus," according to UF research professor Peter Frederick in 2016. "Current levels range between 100 and 300 ppb."

Once inside the marshes, phosphorus boosts the growth of cattails, which thrive as dense mats that crowd out various wildlife. It also prevents the growth of periphyton, a yellowish-brown spongy material that provides food for aquatic insects, frogs, fish, and snails. This disruption causes harm because periphyton, when it decays, produces a soil called marl, one of two primary types in the Everglades.

Too much phosphorus also pollutes the water—which is already dirtied and contaminated by runoff, sewage, animal and human waste, and a host of other industrial-age pollutants. Of course, phosphorus also feeds algae.

So, given these problems, why is sugar even grown in a region lacking in phosphorus? After all, in the early days of dredging, reclaimed muck in the Everglades agricultural area was once considered "black gold"—an ideal soil for growing winter vegetables, not sugar cane.

Sunny, warm days and cheap land provide a partial explanation for the change. Cold War politics played a big role, too. In 1959 bearded revolutionary Fidel Castro and his ragtag guerilla army seized political control of Cuba and installed a pro–Soviet Union, authoritarian government. This takeover spurred a mass exodus of hundreds of thousands of anti-Castro Cubans, most of whom resettled in the greater Miami area.

Among those ex-pats were two young men, the Fanjul brothers, Alfie and Jose, also known as Pepe. As members of a powerful sugar-growing family in Cuba, they feared Castro's capitalist-hating militia and fled the island. After resettling in Florida, the brothers bought tens of thousands of acres of land in the Everglades and set up sugar cane operations.

They worked hard and prospered. But it was also their good fortune—and that of other American sugar growers—that the U.S. Congress was steamed up over Cuba's cuddling up to the hated Reds of Russia. In 1960 Congress decided to teach Cuba a lesson of some sort by imposing an embargo on importation of the island's sugar. This, of course, was a boon for American producers.

Getting help from the federal government wasn't something new for sugar farmers. The first time assistance arrived was during the Great Depression, when Congress placed an import quota on world sugar to help protect domestic growers. There's been more help since then, including a complicated federal sugar program at work today. Under it, the U.S. government guarantees American sugar producers a minimum price for their crops that's often much higher than the world market price. It can do this because it controls how much sugar can be imported from some forty countries. The U.S. Department of Agriculture also buys surplus domestic sugar to keep prices high and converts some of it into ethanol at taxpayer expense.

Though some Florida sugar growers may growl when they hear critics describe the program as a subsidy, or corporate welfare, it is nonetheless true that U.S. taxpayers' money makes life easier for them. Healthy profits also make it possible for Florida's Big Sugar titans to lobby all levels of government to ward off attempts at environmental regulation, tax reform, and real estate purchases not to their liking.

Fixing the Nasty Flow

But even these powerful people could not fend off the growing public outrage over the never-ending news reports and complaints from environmental groups that the Everglades was sick and that Big Sugar was allegedly one of the biggest culprits for its demise.

A tipping point arrived in 1982, when a pro-environment governor and former developer and rancher, Bob Graham, launched a "Save Our Rivers" program, which empowered the five water management districts to buy 1.7 million acres of land to acquire and protect water sources. A year later he promoted a "Save Our Everglades" action plan.

But the health of the Everglades basin kept nosediving. By 1987, you didn't need to be a scientist to recognize that something was wrong with the dying seagrass in Florida Bay, the Glades' terminal river destination. There, a much-reduced volume of water was discharging a witch's brew of pesticides, insecticides, and nutrients.

Dexter Seeks Justice

Much of the bad news focused on how sugar cane farms were polluting the Everglades, and a Miami-based U.S. attorney named Dexter Lehtinen didn't like it. As a boy, he'd grown up in the farming community of Homestead and had fished in the nearby Everglades. In adulthood, he still had a fondness for the Swamp.

Lehtinen, a battle-scarred Vietnam veteran, was also unafraid of a fight. Without obtaining permission from his Reagan-era bosses in Washington, D.C., he filed a federal lawsuit against the South Florida WMD and DEP for failing to prevent water pollution in the Everglades National Park and the Loxahatchee National Wildlife Reserve.

His action sparked controversy and made headlines. The water district countersued, claiming that since the Army Corps of Engineers had built the water control infrastructure, any problems in the Everglades were their fault. A flurry of legal actions followed. In 1991 it all came to an abrupt halt, however, when Bob Graham's successor, Lawton Chiles, strode into the Miami courtroom where the lawsuit was being heard. Up until that moment, recalls Lieutenant Governor Buddy MacKay, who accompanied Chiles to the courthouse that day, the state regulatory agency had spent $6 million defending itself by arguing, "the Everglades waters were not polluted, and . . . even if the Everglades were polluted, it was not the state's fault."

But this argument withered after the sudden appearance of the governor, a self-described "He-Coon." Chiles, a folksy southern populist Democrat from Lakeland, years before had trekked the length of Florida—1,003 miles—to gain public recognition and votes. His walk into the Miami courtroom also made news and history.

To the astonishment of everyone in the courtroom, the sly He-Coon asked for and was granted permission to speak. Chiles readily admitted the Everglades was indeed polluted. "I am ready to stipulate today that water is dirty," he said. And now the question was, as Chiles put it: "Your Honor . . . how do we get clean water? What is the fastest way to do that? I am here and I brought my sword. I want to find out who I can give that sword to . . . and have our troops start the reparation."

Two years after the courtroom drama, all parties signed off on a legal agreement to end the dispute without anyone admitting fault. The federal and state governments also promised to work together to find a common solution to the Everglades mess. In 1994 they did reach an accord, which, among other things, required both the state and the sugar industry to pay for the cleanup.

The state legislature followed up with the Everglades Forever Act, which set a goal of achieving the same standard for water quality for the Everglades that was stipulated in the consent decree. The act also allowed the South Florida WMD to impose taxes on farmers and other landowners in the sixteen-county district and use the revenues to purchase land around the Everglades as a buffer against development and to build a network of artificial marshes to filter out pollutants from the water before it headed south to the Everglades National Park. Finally, the agreement required farmers to adopt best-management practices to voluntarily cut back on the pollution they generate on their farms.

In 1999 the Corps, which was already busy repairing the Kissimmee River basin, had by now also come up with a "restudy" of the Everglades system—at the behest of Congress—which included suggestions from a special governor's commission. This one was designed to bring back the natural ecosystem in South Florida without sacrificing flood control or the supply of water to cities and farms.

Reaction to the study revealed that the environmental, scientific, economic, and political issues related to fixing the Everglades were going to be complex, costly, and contentious. Scores of preliminary studies and reviews managed to annoy all the stakeholders. How the cleanup should be carried out was one big source of contention.

Another was water allocation. At the federal level, for instance, champions of the Everglades wanted any new restoration programs to focus on ecological repair. However, many Florida legislators viewed a restored Everglades as a way to increase the water supply for South Florida to enable future development. This angle prompted one Ohio lawmaker in Congress to wonder why he should support economic development in Florida, which was actively recruiting people from his home state and thus hurting it. Many agriculturalists, meanwhile, wanted to keep the water spigot open for themselves. And scientists at the Everglades National Park warned that the park wasn't going to get all the water it needed. Many environmentalists opposed any kind of compromise between the warring parties that didn't put the rehabilitation of the Everglades environment front and center.

At last, federal and state officials pondered all these considerations and came up with a massive eco-recovery plan for the entire Everglades basin to atone for a century of abuse. On December 11, 2000, President Bill Clinton signed the bipartisan Comprehensive Everglades Restoration Plan (CERP) that authorized a mammoth federal-state project expected to take thirty years to finish. Billed as the biggest restoration undertaking in history, it had an initial price tag of $7.8 billion, which later rose to $20.5 billion, with the state and the federal governments splitting the costs.

The broad goals of CERP called for restoring to the Everglades the "right quantity, quality, timing, and distribution" of fresh water. These characteristics are inseparable from one another. Timed releases of water from Lake Okeechobee are vital to the downstream flora and fauna that had adapted to South Florida's two seasons, which are wet and dry.

Overbearing gushes of discharged water roaring all at once into the coastal estuaries can be devastating to its native marine life. And the nutrients the discharged waters carry do even more damage when algal blooms explode.

Apparently, CERP's managers and engineers were expected to possess godlike creative powers. For example, they were authorized to provide *additional* supplies of fresh water for future development in South and Central Florida, and they had to keep agricultural interests happy while making sure all those canals, rivers, and lakes had enough water to support Florida's ever-increasing number of boaters. Meanwhile, they had to maintain safe navigation through the Everglades from coast to coast and sufficiently hydrate the Everglades National Park.

Water managers were also expected to clean up the dirty water to protect the environment and make sure Lake Okeechobee didn't flood. Achieving both the goals seems a paradox. In part, that's because of the controversial practice of back pumping.

Back Pumping Blues

Most of the time sugar cane and vegetable growers use various canals to draw water southward from the lake to irrigate their crops downstream in the Everglades agricultural area.

But when intense rainfall arrives, South Florida WMD managers pump water out of the farming area and send it north back into Lake Okeechobee. This practice helps protect wildlife, but it also benefits sugar growers because sugar cane crops are especially vulnerable when the soil isn't dry enough.

Thus, Lake Okeechobee became a reservoir for back-pumped water, which sugar growers could use later during periods of dry weather to irrigate their crops.

However, back-pumped water can be polluted with fertilizers, pesticides, and other contaminants that cause fish kills and trigger algae blooms that endanger drinking water supplies.

Whenever lake water levels are rising from heavy rain, lake water managers start to worry about dam security and flooding. If the lake level becomes dangerously high, they are forced to discharge water with all its foul pollutants via the St. Lucie and the Caloosahatchee canals to the coasts, enraging the residents living there.

In 2002 the attorney-packed environmental group Earthjustice temporarily won a federal lawsuit seeking to stop back pumping. The district, they alleged, violated the Clean Water Act, because it didn't get a permit before it transferred the backpumped water.

But the South Florida WMD continued the practice anyway, while it appealed the decision, claiming the danger of flooding left the district no other choice. On January 18, 2017, a federal appeals court overturned the earlier decision. So back pumping continues when needed.

Could Disaster Happen Again?

Ride around Lake Okeechobee, and you understand why there's concern for flooding. The dike rises ominously above everything in the region. Here and there you see settlements of the poor—clusters of ramshackle, dilapidated mobile homes on tiny lots semi-shaded by royal palms with yard signs extolling "Bubba's Mattresses" and "Lake Property for Sale." In some places, these flimsy, blow-away dwellings are huddled together halfway up the embankment of the dike itself. Anyone with the foggiest historical sense of the deadly floods of the 1920s can't help but wonder why these homes are allowed to be there at all.

Even in Clewiston—America's "Sweetest Town," located on "Sugarland Highway," a company town, built by U.S. Sugar right next to the lake in the 1920s—there's an air of uneasiness when lake waters are rising. An outdoor sign across the street from the Clewiston Inn keeps track of how high the lake is getting. "Once it hits about sixteen feet, our insurance companies tell us we've got to get flood insurance," says a tall, dark-haired woman who owns a local diner.

Eighteen feet is about the limit. That's when the engineers reach for discharge controls.

The Corps, however, is now embarking on a massive dam strengthening project that should provide an extra layer of security.

When hurricanes blow in from the Caribbean, the Atlantic, or the gulf, Floridians living on the coasts hurry inland. Not so in Lake Okeechobee country. The evacuation signs posted in the prosperous-looking town of Okeechobee, situated on the lake's northern shore, explain what everyone is supposed to do when a mighty storm blows inland: evacuate to the north. In other words, get away from Lake Okeechobee!

Anger on the Coasts

Residents living along the Indian River Lagoon and the Caloosahatchee have long understood the dangers of floods. What they don't understand is why discharged Lake Okeechobee waters have to be so polluted. Many are quick to blame farmers in the EAA.

According to Manley Fuller, president and CEO of the Florida Wildlife Federation, Big Sugar—which in Florida refers to the state's largest sugar companies, U.S. Sugar, Florida Crystals, and the Sugar Cane Growers Cooperative—contributes 62 percent of the total phosphorus load that ends up in Lake Okeechobee and the St. Lucie and Caloosahatchee waterways. Vegetable farmers contribute more on top of that.

Others, including scientists, say Fuller's figure is too high, and instead argue that nutrient-rich pollution from urban Orlando, flowing south in the Kissimmee River, supplies the lion's share of phosphorus and nitrogen to the Lake Okeechobee region.

Sugar growers don't deny they produce pollution, and many of them have willingly implemented programs to reduce their phosphorus footprint. Big Sugar is also quick to use its huge political clout to tamp down any reforms that would impose mandatory requirements on farmers to clean up their act. Such things, argue Big Sugar's public relations pros, are job-killing, anti-free market, land-grabbing, and an affront to American farmers who feed this nation!

Earthjustice's David Guest is cynical about such rhetoric. A veteran litigator, he's gruff, imposing, and speaks like an angry poet if he gets revved up.

"Listening to Big Ag guys, you hear so many who are indignant," he says. "But there's an inconsistency between the free-market philosophy they espouse when they get so many free passes in terms of subsidies and then not have to clean up their own pollution. These folks have a view that what they do is so virtuous that they should get free passes. They think they are the modern nobility that is above the law."

State pollution regulations for agriculture are hardly onerous. For the most part, Florida farmers—whether dairy farmers in the Suwannee River region, EAA sugar farmers, or citrus growers or others—need to demonstrate they are complying with state standards to reduce water pollution in their agriculture operations or adopt one of the state's numerous best management practices (BMPs). In the past, much of what they did was largely self-monitoring.

By 2016 only about 50 percent of farmers statewide had signed up for the state plan, according to Steve Dwinell, director of water policy with the Florida Department of Agriculture Consumer Services (DACS).

That percentage annoys Whitey Markle, chair of the Suwanee-St. Johns Group Sierra Club, who is frustrated there aren't stronger regulation and prevention of pollution caused by agriculture. "DACS has continuously made excuses for not monitoring and not staffing," he complains. "The government is a great big, slow-moving machine, by legislative design. Apparently, the current tactic is to continue the age-old stalling technique."

At least Dwinell might agree with Markle on one thing. The use of BMPs, he says, by themselves will not allow agricultural interests to reach the goals set by the state, "so we need something else."

Sugar industry spokespersons, however, say their people are not about to accept full responsibility for pollution in the Everglades; others are also to blame. Just look at all those subdivisions with septic tanks along the Indian River Lagoon in the east and more of the same near Ft. Myers on the Caloosahatchee, they argue. Seepage from them mixes with lawn fertilizers, and wastewater and all are flushed into the estuaries. They also fault dairy farms and cattle operations north of the lake that contribute animal waste teeming with nitrates.

Brian Lapointe, a burly, tanned research professor at the Florida Atlantic University (FAU) Harbor Branch Oceanographic Institute in Ft. Pierce, thinks there's some truth to these arguments. Relying on his own research, he's concluded that more than 95 percent of the water that ends up in Lake Okeechobee originates north of the lake, which includes the upper Kissimmee River basin, an area exposed to all sorts of urban contaminants from greater Orlando.

Lapointe also points out that phosphorus wasn't the only algae food found in the Indian River Lagoon; there was also nitrogen, whose origin is often human and animal waste. After the big, and controversial, blowout from Lake Okeechobee to the St. Lucie in 2016, he led a study team to determine the source of a huge algal bloom—this one a brown tide caused by a different kind of algae than the well-known blue-green variety—in the northern Indian River Lagoon. The most abundant and dangerous substance found in the putrid waters that day was wastewater nitrogen, which indicates human sewage.

"We can tell whether that nitrogen is from fertilizer or from sewage by collecting algae and looking at the ratio of N15 to N14 isotopes [slightly different forms of nitrogen], and what we found was that wastewater nitrogen was the major source of nitrogen fueling these algae blooms in Florida's Indian River Lagoon—the whole 156-mile stretch," says Lapointe.

This isn't to say that pollution from Lake Okeechobee played no role. Lapointe's research team detected isotopes "that told us that the water that came from Lake Okeechobee was carrying seed algae." When this seed-saturated flow entered the St. Lucie Estuary, it feasted on wastewater nitrogen and phosphorus that were already there and then grew quickly and abundantly and became more toxic. Lapointe suspects that much of this nutrient-rich wastewater seeped from some of the tens of thousands of septic tanks in Martin and St. Lucie Counties.

In fact, he argues, "phosphorus concentrations were up to 600 percent higher in the North Fork of the St. Lucie Estuary than in what was coming in from Lake Okeechobee."

Most of the polluted water from farmers, he argues, flows south of Lake Okeechobee. Very little, if any at all, he says, ever makes it to the northern portion of the beloved lagoon.

Begging to Differ

Not everyone agrees with Lapointe's analysis. Gary Goforth, chief consulting engineer of the Everglades construction project, for instance, believes the septic tank issue has been used as a "red herring during the discussion of remedies for reducing the damaging lake discharges to the estuaries." In the process, he says, it diverts and deflects "attention away from the most important problem—reducing lake discharges and stormwater runoff from agricultural lands."

Goforth, a tall, athletic-looking man, agrees that any septic tanks that contribute loading to the estuary need to be fixed; however, he argues, they aren't the biggest problem. Instead, he turns to a 2013 DEP study that he says found nitrogen loading from septic tanks to the estuary to be less than 10 percent of the total inflows to the estuary.

There's also the problem of waste. "On average," he says, his voice rising on this point, "242 billion gallons of water from Lake Okeechobee are wasted annually by being sent to the Atlantic Ocean via the St. Lucie Canal and to the Gulf of Mexico via the Caloosahatchee River. If additional water is allowed to flow from Lake Okeechobee to the Everglades—instead of being discharged to the estuaries—it could help to combat the effects of sea level rise and slow the advance of salt water intrusion that could contaminate drinking-water wells."

Proving whose science is right is always hard. Even tougher, perhaps, is trying to figure whose research and engineering projects should receive tens of millions of dollars in funding to address the problems.

Partisan politics in Florida, as always, often goes into hyper mode when the issue is Everglades restoration. Finger pointing abounds. Governor Scott, for instance, chose to blame President Barack Obama and the federal government for the 2016 algal catastrophe in Martin County. "Although the President has failed to do what is needed to address this growing issue," he said at a press conference, "the State of Florida will devote every available resource to find solutions for the families and businesses in this area."

Scott's critics shot back that his attacks were misplaced. For one thing, they replied, the dike's overdue repairs depended on funding priorities of Congress. They also reminded Scott that his own administration had cut the budgets of state agencies responsible for pollution enforcement and that he had also opposed federal efforts to make Florida's water-quality standards stronger.

"This year, the Legislature passed, and Scott quickly signed, a water-policy bill that essentially allows major agricultural producers—whose fertilizer use is a top contributor to water pollution—to police themselves," the *Orlando Sentinel* editorialized on July 17, 2017. "Last year Scott, lawmakers and the governor's appointees on the South Florida Water Management District let an option expire on purchasing sugar-industry-owned land south of Lake Okeechobee so that more lake water could be treated and sent south into the Everglades, rather than routed to the coasts."

Everglades Fix-It Man

Florida Senate president Joe Negron, like Scott, belongs to the Republican Party, many of whose members oppose state purchases of private property. In fact, many Republicans have publicly urged Florida to sell "surplus" state lands Floridians already own.

Negron, however, also represents the people of Stuart, where the nearby Indian River Lagoon gets drubbed over and over again by algae invasions. In the summer of 2016, they were outraged more than ever and demanded action.

And so, Negron acted. On August 9 of that year, he proposed a bill that would authorize adding 120 billion gallons in water storage south of Lake Okeechobee to take the pressure off the estuaries and clean up the toxic water before sending it to rehydrate the Everglades properly.

Thousands of residents in Florida's Treasure Coast, along with environmentalists across the state, now considered Negron "The Man." At long last, they believed, a politician—even if he did have ties to the sugar industry—was actually doing something about the Everglades.

Not all his constituency was pleased. Many residents in Pahokee and other rural areas on the southwest side of the lake, in fact, were alarmed, especially if the muscle power of eminent domain was going to be deployed by the state to force unwilling landowners to sell their land for the greater good.

Almost overnight, a Big Sugar public relations campaign materialized. New grassroots organizations hatched out of nowhere, bearing names like Glades Lives Matter. Hundreds of jobs would be lost, they warned. Land grab!

But there was also genuine concern expressed from others about the impact a big land purchase could have on the poor. "What about the people?" asked Priscilla Taylor, a former state representative and Palm Beach County commissioner. "Not only will the 1,000-plus workers be out of work, but the banks, grocers and other businesses will be negatively affected as well."

South Florida WMD officials jumped into the fray, as well. Pete Antonacci, the water district's new executive director, was not an experienced water professional. Instead, his background included being a former lobbyist and general counsel for the governor. Nonetheless, he, along with Senator Marco Rubio, the governor, representatives of Big

Sugar, and water district officials ganged up to oppose Negron's idea. All agreed a reservoir wasn't needed in the EAA. Go north of the lake, they insisted, and take advantage of restoration plans already under way or slated for that area. Clean up the water before it gets into Lake Okeechobee; doesn't that make sense?

In late February 2017, water district engineers spoke up, too, resurrecting earlier plans that called for using aquifer storage and recovery systems, which involve pumping water into underground storage areas and retrieving it when needed.

Many of Negron's supporters suspected that a lot of this ruckus was Big Sugar's doing. After all, just eight years earlier, then-Governor Charlie Crist brokered a deal in which U.S. Sugar had agreed to sell 187,000 acres south of Lake Okeechobee to the state for the very purpose of saving the Everglades. But the Great Recession dried up funds, and Governor Scott walked back from the deal. However, the state still had an option to buy U.S. Sugar's land for $1.75 billion, if it did so before December 2016, when the contract expired. To many it seemed like a good backup plan, should Negron's efforts fail. But now U.S. Sugar didn't want to sell. After all, land values have gone up since the recession.

The reservoir-south forces soldiered on. Yes, they responded, a water cleanup was needed in the north, but it wouldn't reduce phosphorus loading in the sugar cane fields, would it? And it wouldn't provide needed water in the south.

On June 7, Negron's forces became even more suspicious when South Florida WMD's governing board voted unanimously to plunge ahead with a new plan to inject billions of gallons of fresh water about three thousand feet into the earth—the boulder zone—before it discharges into Lake Okeechobee from the north. The board, and its supporters, claimed this method was a cheaper way to keep polluted water out of St. Lucie and the Caloosahatchee.

Critics of this idea, including the Corps of Engineers and environmentalists, were put off by this deep-well injection idea. Why would anybody sink fresh water so deep into the earth that they'd never be able to retrieve it? What if some of the untreated water made its way through limestone cracks and contaminated groundwater? The whole thing sounded eerily like something the dredge-and-drain bunch would have promoted a century ago.

In February, State Senator David Simmons, from Longwood, complicated the ongoing north-south feud with a proposal to fortify the Hoover Dike by raising its walls by two feet to trap more water, and thereby giving some relief to the coastal communities by reducing the size of the discharges. Governor Scott favored the idea, too.

Objections arrived faster than a FedEx delivery. Some legislators complained that the dam overhaul would be too expensive and would deflect energy and discussion away from Negron's push. Lake Okeechobee biologists claimed raising lake levels would flood out snail kites, native grasses, and other vegetation that provide nurseries for life in the lake.

As the debate intensified in the spring of 2017, Wendy Graham, director of the University of Florida's Water Institute, wanted to set the record straight. She told this writer that a UF 2013 report that was being cited by all sorts of belligerents duking it out over the Everglades wasn't a study at all, but rather a survey—a "synthesis"—conducted by UF scientists who reviewed major previous research projects involving Glades' water issues existing from the 1990s. The one consistent finding the scientific team came across, says Graham, was that a "massive" amount of water is needed to fix the Everglades.

Arguing whether the north or south is the best location for a reservoir may be missing a bigger point. The UF analysis, she insists, reveals, "Reservoirs both in the north *and* south are necessary, and they need to be built as fast as possible."

A New Deal?

By May 10, 2017, Florida lawmakers had finally agreed on a new bill, which Scott signed into law. It calls for both the state and the Army Corps each to contribute $800 million to pay for a new 78.2-billion-gallon reservoir. That is two-thirds the size of the huge pool Negron had initially wanted.

Since then, however, there's been an outbreak of bickering about the size, location, and capacity of a new reservoir among environmentalists, legislators, and officials of the South Florida WMD, which now had a new executive director, Ernie Marks, who unlike his predecessor had twenty years of public service in natural resource management.

Before any digging occurs, however, Congress must authorize any federal undertaking, and President Donald Trump has to sign off. Whatever the final cleanup and restoration of the Everglades turns out to be, it will be a daunting task.

"The size, the scale of this watershed is staggering," says water consultant Rich Budell, former director of the Office of Agricultural Water Policy with the Florida Department of Agriculture and Consumer Services. He also has coauthored much of Florida's important water legislation, such as the Florida Watershed Restoration Act and the Lake Okeechobee Protection Act.

"It'll take years to design and permit these facilities, and then it'll take years to see improvements," he says. "This could be hard to do because the public is accustomed to quick technological fixes. The challenge is to have the wherewithal to stay on course and maintain the effort for generations."

But can the Everglades wait that long? In November 2017 the International Union for Conservation of Nature (IUCN) issued its conservation outlook assessment of 241 World Heritage natural sites. Seven percent of these locations received a "critical" rating. The only American site to receive such a dire outlook was the Everglades National Park. According to the report, "major issues with water quantity, quality, distribution and timing, invasive species and climate change are creating over-riding impacts to the system, and the continued deteriorating trend of so many values puts the park's World Heritage values in a critical situation."

More abuse soon piled on. In the summer of 2018, the twin juggernaut of algal blooms and red tide struck Florida, causing statewide outrage. The Everglades, Lake Okeechobee, and Florida's sickening waterways were hotter political topics than ever.

But were politicians ready at long last to take action?

They might do well to consider another solution to the Everglades mess—a wide-ranging one that gets scant attention. It comes from W. Hodding Carter, author of *Stolen Water: Saving the Everglades from Its Friends, Foes, and Florida.* "I know what we should do if we really want to restore it," he argues. "This won't take long. It's really very easy: the wackos have been suggesting it for decades. Do the one, and only one, thing that works in the country when you want to preserve an environment. Make the entire historic Everglades a national park—from Lake

Okeechobee on down to Florida Bay. Get rid of this half-assed solution called the Comprehensive Plan."

By the time that could ever happen, something that is now creeping in from the sea could make a mockery of all the time, energy, and money now being spent to save the Everglades. In fact, the Everglades may just cease to exist.

Blame it on sea level rise.

11

Flooding Time Again

FLORIDA IS AT "ground zero for climate change." This is what Senator Sheldon Whitehouse, a Rhode Island Democrat, testified before a special U.S. Senate committee in June 2014. Geography 101 explains why the senator's case was easy to understand. For one thing, Florida is a peninsula surrounded on three sides by seawater. Ancient seawater lies beneath the aquifers as well. The karst rock underneath Florida's soil is fragile and filled with billions of holes, which can't be plugged. There's no way to keep seawater out of them.

Much of Florida's coastal landmass is only inches above sea level. The Sunshine State has more than twelve hundred miles of coastline—that's more than any other state except Alaska. About 80 percent of Florida's population lives within twenty miles of a beach. Residents get most of their water from nearby wells, some of which are already impacted by sea level rise.

But is climate change really happening? The overwhelming majority of climate scientists around the world think so, and they also are convinced it is largely human-caused. Their assessment is expressed by the 195-nation United Nations Intergovernmental Panel on Climate Change (IPCC). According to the panel's 2014 report, based on the contributions of thousands of scientists, "Human influence on the climate system is clear, and recent anthropogenic emissions of greenhouse gases are the highest in history. Recent climate changes have had widespread impacts on human and natural systems. Warming of the climate system is unequivocal, and since the 1950s, many of the observed changes are unprecedented over decades to millennia. The atmosphere and ocean have warmed, the amounts of snow and ice have diminished, and sea level has risen."

The federal government's own premier report, "The National Climate Assessment" (NCA), prepared by more than three hundred experts and guided by a sixty-member Federal Advisory Committee, concludes "that the evidence of human-induced climate change continues to strengthen and that impacts are increasing across the country."

Several Florida coastal governments are worried enough to do their own research. The Miami–Dade County Task Force on Climate Change, for example, predicts: "Miami–Dade County as we know it will significantly change with a 3–4 foot sea level rise. Spring high tides would be at about +6 to 7 feet; freshwater resources would be gone; the Everglades would be inundated on the west side of Miami–Dade County; the barrier islands would be largely inundated; storm surges would be devastating; [and] landfill sites would be exposed to erosion[,] contaminating marine and coastal environments." The Village Council of Pinecrest, on the outskirts of South Miami, has changed its comprehensive development plan to take climate change into account.

Nor is this concern limited to that of scientists and governments. The international business community is worried, too. According to *The Global Risks Report 2016,* published by the World Economic Forum, "After its presence in the top five most impactful risks for the past three years, the failure of climate change mitigation and adaptation has risen to the top and is perceived in 2016 as the most impactful risk for the years to come, ahead of weapons of mass destruction, ranking 2nd, and water crises, ranking 3rd."

The U.S. military is also worried that sea level rise is likely to damage 1,774 of its installations worldwide. Florida has 21 military sites, and more than half of them are close to the coast. NASA officials, meanwhile, are concerned about their agency's own infrastructure staying dry.

An Old Story

Although controversy over climate change has flared up over the past few decades, it's really old news. Seawater has always been a threat, especially to Florida. Over the ages, it has ebbed and flowed across the state, according to the rise and fall of ice ages and other natural forces. Today, it surrounds most of the state. Salty water also lurks beneath the underground reservoirs of fresh water known as aquifers. This happens because salt water—having dissolved salts—is denser and thus heavier, allowing fresh water to float above it. Thus, seawater is always there deep below, threatening to rise, if the pressure of fresh water goes away. Pump too much fresh water from the aquifer, and the salt water underneath starts to rise.

What's happening today is different. Sea level rise is happening faster than ever before thanks to a worsening of a global greenhouse effect. In 1824 French physicist Joseph Fourier figured out that a natural greenhouse effect was always at work. Plants and trees, after all, give off carbon dioxide, the gas in the atmosphere most responsible for trapping heat trying to escape the earth's surface. But until human activities became disruptive, nature had achieved a balance: The amount of energy striking the earth from the sun equaled the amount "radiated" or emitted in waves of energy back up to space.

Almost four decades after Fourier's discovery, Irish physicist John Tyndall realized that water vapor and other gases created a gaseous blanket around the earth that trapped some of that radiated heat. In 1896 Swedish chemist Svante Arrhenius discovered that the massive burning of coal brought on by the Industrial Revolution was adding heat-trapping power to the vast celestial, artificial greenhouse in the sky. Another Swede, Knut Angstrom, soon discovered that carbon dioxide actually absorbs infrared heat, increasing the warming effect even more.

In 1938 British engineer Guy Callendar collected data from 147 weather stations around the globe and concluded that the earth's annual temperatures had risen since the nineteenth century. He also found this temperature rise was accompanied by an increase in carbon dioxide in the atmosphere. Was it a correlation or was it cause and effect? At the time, his colleagues were not too interested in finding out.

Today's scientists are, and they've concluded that anthropogenic, or human-caused, air pollution is amplifying the warming of the planet and causing climate change.

Average temperatures across the United States have gone up 1.3°F to 1.9°F since 1895, the first year scientists began keeping such records. Most of the noticeable uptick, however, has been since 1970. In 2018 scientists noted that the previous ten years had been the hottest since temperatures have been recorded. Worse yet, they expect temperatures in much of the United States to keep rising.

"Anthropogenic greenhouse gas emissions have increased since the pre-industrial era driven largely by economic and population growth," according to the IPCC report. "From 2000 to 2010 emissions were the highest in history. Historical emissions have driven atmospheric concentrations of carbon dioxide, methane and nitrous oxide to levels that are unprecedented in at least the last 800,000 years, leading to an uptake of energy by the climate system."

That's not all that's going up. Over the past 130 years, sea level has risen by about eight inches. Some scientists expect seas to rise another one to four feet by 2100. The National Oceanic and Atmospheric Administration has a broad range of projections, and one of them is that seas could rise as much as eight feet by the end of the century.

Melting land-based ice sheets and glaciers account for much of the rising sea levels. But warming temperatures are also heating up the seas and causing their water molecules to expand. This expansion causes water levels to rise. Boiling water spills upward and over the pot for the same reason.

If these trends continue, say the vast majority of climatologists, Florida may be the most vulnerable of all the states to climate change.

The scientific research is replete with disturbing predictions: flooded streets, salt water intrusion into aquifers, interrupted water supplies, the northward spread of tropical diseases—such as malaria and dengue fever, along with marine-borne illnesses and shellfish poisoning—an increase in algae blooms, unbearable heat, loss of farmland, the bleaching of Florida's reefs, loss of wetlands and estuaries, loss of habitats, massive relocation of infrastructure, abandoned neighborhoods and business regions, a cooler Gulf Stream, stronger hurricanes, an influx of Caribbean refugees, and a loss of tourism.

In the past fifty years, sea levels along Florida's coasts have risen five to eight inches. This seemingly small amount, however, has been accumulating, and it doesn't take a lot of sea rise to get noticed along Florida's coasts, which are scarcely above sea level in the first place.

One Floridian keeping an eye on sea level rise is Andrea Dutton, a carbonate geochemist and sedimentologist with the Florida Climate Institute at the University of Florida. Dutton is a woman on a mission. She's a scientist foremost, but she also sees herself as an educator who feels compelled to help others understand the big change that is coming. That's why she agreed to be one of the principal speakers at the March for Science in Gainesville in March 2017.

There, she told the audience of her field work, which consists of a lot of globe-trotting to study the behavior of sea level and ice sheets during interglacial periods. She's conducted research in many exotic places and seen many things, but one of the most revelatory experiences happened while she was in the Seychelles Islands in the West Indian Ocean.

"We were interested in what was going on about twenty thousand years ago, when the climate was warm like today, if not a little warmer, especially at the poles, where we have these large ice sheets," she later told this writer. "We wanted to see how much ice melted in the past and how high the sea level rose."

The research team used a dating technique that measured the age of uranium isotopes absorbed into the coral reefs there. Some of those reefs now stand eight meters—or more than twenty-five feet—above present sea level.

"This was really astounding to me. I didn't expect it to be that high. I sat down on a rock and asked myself how am I going to explain this? People are not going to like this at all. It really implies that sea level rose a lot with just a little bit of higher temperature."

The only way those coral reefs could have been swamped in the past is that the ice sheets had melted. Part of what Dutton and her colleagues do is to see if the past climate changes help indicate what will occur in the future.

So far, it doesn't look promising. Many scientists say even if the world's air pollution magically ended today, there's already so much of it baked into the environment that warming will continue for many centuries, or longer.

Most climatologists are certain that climate change is happening, but they are uncertain about the effects of these changes. New research from the Jet Propulsion Lab, for instance, suggests the earth itself may be operating like a huge sponge and absorbing much of the moisture coming from melting ice sheets and thus keeping it out of the seas—for now—and slowing down the rate of the sea rise.

Even if the JPL is right, sea rise around Florida isn't slowing down or stabilizing. Consider low-lying Miami and its myriad of canals and floodgates constructed more than a half century ago. Until recently all that South Florida water engineers had to do to get rid of stormwater after a downpour was to open the floodgates and let the stormwater flush out to sea. Today, however, in some places during high tides, stormwater can't evacuate because the sea level is *higher.*

There's yet another recent twist—just one of many—in the story of sea level rise. In August 2017 Dutton and Arnoldo Valle-Levinson, a UF professor of civil and coastal engineering sciences in the College of Engineering, published a study online in *Geophysical Research Letters* that shows sea rise in the southeastern United States was six times greater than the worldwide sea level rise triggered by human-caused global warming. The study's findings suggest that naturally occurring, localized "hot spots" are responsible. They are caused by the combined effects of El Niño (which throws more water against the coasts) and the North Atlantic Oscillation—a periodic seesaw fluctuation in atmospheric pressure at sea level in the North Atlantic Ocean. Together they magnify the steady sea level rise already under way.

"If you look at the tide gauges that are situated all along the Florida coastline," says Dutton, "you'll see that all of them are reporting a rate of sea rise that is . . . faster than the global average—that's not the best of news."

In the past, these hot spots came and went. How they will behave in the future alongside climate change isn't clear. But one thing is certain: steady anthropogenic sea level rise shows no signs of abating.

In coming years, unless major reconfigurations are made to the South Florida canals, ocean water will flood perilously inland. If the most recent National Climate Assessment predictions prove accurate, by the end of the century, low-lying areas, such as Centennial Park in downtown Ft. Myers, could be underwater. By 2100, sea level rise could be as high as 6.6 feet, inundating much of Miami–Dade County. If nothing is done to slow down this rise, much of South Florida may be underwater by 2200.

"To consider the risk in present investments is beyond sobering," writes Harold R. Wanless, professor and chair of geology at the University of Miami, and a leading authority on sea level rise in Florida. "By the middle of this century most of the barrier islands of south Florida and the world will be abandoned and the people relocated, while low areas such as Sweetwater and Hialeah bordering the Everglades will be frequently flooded and increasingly difficult places to live. Florida will start to lose its freshwater resources, its infrastructure will begin to fail, and the risk of catastrophic storm surges and hurricane flooding will increase."

A Never-Ending Threat to Florida Fresh Water

Climate change will significantly impact the sustainability of water supplies in the coming decades, if not centuries. A 2010 analysis performed by consulting firm Tetra Tech for the Natural Resources Defense Council (NRDC) "examined the effects of global warming on water supply and demand in the contiguous United States. The study found that more than 1,100 counties—one-third of all counties in the lower 48—will face higher risks of water shortages by mid-century as the result of global warming." Florida is one of the fourteen states that is likely to run a "high risk to water sustainability" as demand outstrips supply.

Seawater intrusion is one of the many effects associated with climate change expected to threaten Florida fresh water. Today, salt water is moving farther inland than ever before. Not long ago, it had crept nearly six miles inland in the Ft. Lauderdale area, forcing water management officials to abandon wells and move inland for new sources. Hallandale Beach, meanwhile, has abandoned six of its eight drinking-water wells. Today its local government is purchasing half of its water from nearby Broward County.

Rising sea levels are slowly covering up barrier islands and altering the water quality of large estuaries such as Tampa Bay, Charlotte Harbor, and the Indian River Lagoon. Seawater is showing up at the headwaters of the Biscayne Aquifer, southern Miami–Dade County's water supply.

There are other hot spots too. "The Pensacola Bay and St. Johns River watersheds and southern Florida from Palm Beach to Miami, the

Florida Keys, Naples, and Fort Myers are especially vulnerable to saltwater intrusion into municipal freshwater supplies," according to the Florida Oceans and Coastal Council, a research group created by the state legislature in 2005.

In addition to sea level rise, say scientists, hotter temperatures are expected to intensify the evaporation of water from the soil and plants into the air. This process could dry out vegetation in some parts of Florida, possibly resulting in drought-like conditions.

Paradoxically, other areas of Florida could get wetter. Since warm air absorbs more moisture, humidity could increase, resulting in increased rainfall and flooding. Some meteorologists anticipate warmer ocean water and increased humidity to fuel stronger hurricanes. Subtropical conditions, meanwhile, could expand northward from South Florida, if warmer temperatures and rainfall continue.

All this uncertainty exists because climate change is an unfolding phenomenon. However, rainfall patterns already seem to be changing. "There is scientific verification that the interior area of Florida has been receiving less rainfall in the past one hundred years," says Tom Champeau, division director of Freshwater Fisheries Management with the Florida Fish and Wildlife Conservation Commission. "When it does rain, it is more frequent in some areas of Florida, but there's also longer periods of drought elsewhere. The length of time that some rivers, lakes, and groundwater levels stay at a low level is greater now than it was in the past. If you're trying to be adaptive to climate change, we have to take this sort of data into consideration with our planning."

The Florida Keys and South Florida have been experiencing less rainfall in recent years, but at the same time parts of Central and North Florida have seen more.

Disturbances in rain cycles could significantly alter how water percolates and travels throughout the state and drains into rivers, streams, and water recharge areas.

Just how severe these changes may be remains unclear. As a 2011 white paper on climate change and Florida's water resources, written by a team of state university researchers, observes, "The uncertainty about how climate change will impact Florida's water resources and its infrastructure creates a challenge for most sectors of Florida's economy. Uncertainty is exacerbated by the strong human imprint on water resources through consumption of fresh water, disposal of wastewater, alterations in land cover, and control of surface water flows."

Calusa Waterkeeper and biologist John Cassani frets that the uncertainty factor is making it harder to carry out scientific inquiry. "Climate drives extra variability that creates great uncertainty," he says. "I think that among scientists there is the idea that the millions of hours we put into research and the criterion we used will not be relevant anymore. It's almost incomprehensible."

Cassani is also convinced the accelerating threat of climate change and sea level rise is "magnified in Florida by a population growth rate of 350,000 people per year while losing at least 75,000 acres of rural land to new intensive development per year."

He also blames insufficient regulatory compliance and oversight, along with the dismantling of Florida's growth management plans, as being largely responsible for the widespread impairment of Florida's waters.

Unfolding Legal Disputes?

The ongoing fresh water battles among humans are likely to increase and intensify, as well. Urban areas could end up struggling with agricultural interests over water even more than they already do. South and Central Florida communities may intensify efforts to tap water resources in the state's water-rich north.

Increased bickering over water is certain to spark legal disputes. Florida judges have long had to resolve disputes over property rights and public domain–based takings by local governments. They can expect to hear new cases over water usage, ecosystems, wetland protection, and many other water-related issues. In the process, Florida's local and state judges will be asked increasingly to find ways to strike a balance between public needs and private landowner rights.

Federal courts may face an increase in lawsuits too, many of them anchored in the Endangered Species and the Clean Water Acts. Public safety is another concern. Take the Army Corps of Engineers, for example, which is responsible for Lake Okeechobee. Holding back sea level rise, while also being accountable for a huge freshwater lake that could drown thousands of Floridians, could prove to be a huge logistics nightmare.

The Cost of Flooding

Coping with sea level is going to be costly. Billions, if not trillions, of dollars will be needed to provide the capital outlay to build new reservoirs, dams, dikes, water pipes, and desalination plants and set up pumping stations to provide fresh water. New pumps, seawalls, and dredging operations will also be required to be resilient against sea rise.

Who pays for all this? Developers? Taxpayers? Tourists? Government? No matter who gets the bill, the damage will be expensive. Currently, many Americans living in high-risk flood plains buy federally backed flood insurance through the National Flood Insurance Program, managed by the Federal Emergency Management Agency (FEMA). However, they must live in participating communities that agree to adopt federal flood mitigation regulations.

"As of November 1, 2016, there are nearly 1.8 million flood insurance policies in Florida, which represents roughly 37 percent of total policies in effect nationwide," according to the Florida Division of Emergency Management. "These policies equate to more than $429.9 billion of insurance coverage."

In July 2017 Florida ranked fourth nationally in receiving money from the federal program, which suffers from a $25 billion deficit. How long will such a program remain if federal debt-reducing measures are undertaken in the coming years?

New climate realities are likely to trigger changes in Florida's laws and ordinances dealing with land use and water. "A successful state response to the challenge of climate and sea level rise changes begins with and cannot be achieved without effective land use planning and zoning," writes Richard Grosso, director of the Environmental and Land Use Law Clinic at the Shepard Broad Law Center, Nova University, in Ft. Lauderdale. Good land use planning, however, may be hard to accomplish in Florida's current deregulatory environment.

You can also expect climate change to impact real estate prices. In fact, a recent Harvard study reveals that "climate gentrification" is already under way in Miami. Simply put, multimillion-dollar mansions along the flood-prone coast are now depreciating in value, as working-class homes on drier land five miles inland are gaining.

The pace of sea level rise may alarm climatologists who do the measuring, but for many people the process seems slow. Millions alive now will never see the great changes. Perhaps that explains why there is little sense of urgency.

Nonetheless, trouble is here now, and more is on the way. So what has been the response from Florida's state leaders?

So far, not much.

Skeptics

Despite decades of international scientific research and almost unanimous agreement among the majority of climatologists that climate change is man-made and destructive, skepticism looms large in Florida.

And that's odd when you consider the Sunshine State's vulnerability to sea level rise. Nonetheless, much of the state is a safe haven for a thriving community of climate change skeptics and nay-sayers, including President Donald Trump, whose Palm Beach mansion, Mar-a-Lago, sits on the Atlantic coast.

Rebuffing the consensus of climatologists has become an article of faith among many of Florida's state representatives. Many of them dismiss climate change as an unproven theory. Others say it is a hoax concocted by scientists wanting research money. A few even charge that talk of climate change is a conspiracy, hatched by central-planning socialists to destroy capitalism.

But what about all that flooding now under way in Miami Beach, Sarasota, Tampa, and other coastal cities? Many skeptics reluctantly accept the evidence, but argue humans are not to blame. Instead, they accuse sunspots or the earth's movement through outer space. Some even assert that carbon dioxide plays no role in the greenhouse effect.

Scoffing at climate change wasn't always an orthodox chunk of conservative ideology. For instance, in 2008 former Republican Speaker of the House Newt Gingrich joined Democrat Nancy Pelosi in a television ad expressing their mutual agreement that the nation "must take action to address climate change."

Former Republican Florida governor Jeb Bush also expressed concern about climate change and urged Floridians to diversify their use of fuels and conserve energy. During his administration, the GOP-dominated legislature passed bills that set up a framework for a market-based cap-and-trade system to cut carbon emissions. Lawmakers also established a special commission to study the effects of global warming.

Bush's successor, Charlie Crist, then a Republican, wasn't a skeptic at all. As governor, he announced that global warming was "one of the most important issues that we will face this century." To show he was serious, he issued executive orders to try to reduce current greenhouse gases by 80 percent of the 1990 levels. He also opposed new coal-fired power plants, pushed for alternative fuels, such as ethanol, and held several climate change summits, featuring speakers from around the world. Crist shared the stage with climate change celebrities, such as Sheryl Crow and action movie star turned California governor Arnold Schwarzenegger. Such actions put him in step with a much-admired British conservative leader, UK prime minister Margaret Thatcher, who had majored in chemistry in college and once called for an international treaty to halt climate change.

Over the decades, however, a much-altered Republican-dominated legislature has undone all of Crist's reforms and cast him as a pariah for the heresy of "believing" in global warming. By 2014 many legislators were indifferent, and even hostile, to calls for Florida to wake up to climate change.

These attitudes so irritated the South Miami city commission, which understood all too well that flooding was happening to its shoreline, that it decided to secede from the state of Florida. In March 2015 the commission passed a resolution, which, in part, reads: "The creation of the 51st state, South Florida, is a necessity for the very survival of the entire southern region of the current state of Florida and this cannot be accomplished by one municipality alone."

The commission's frustration with Governor Rick Scott was hard to miss. "Everybody's afraid of antagonizing the guy who doesn't recognize climate change—I forget his name," said then–Vice Mayor Walter Harris.

Harris was slyly referring to a report from the Florida Center for Investigative Reporting that claims Scott allegedly told employees of state agencies "not to use the terms 'climate change' or 'global warming' in official correspondence."

The governor's office denied Scott ever said this, but the widespread impression that the most vulnerable state to climate change had adopted such a policy made Florida the laughingstock of the nation.

Besides, Scott's public views on climate change were well known. "I've not been convinced that there's any man-made climate change," he said during the 2010 Florida election. "Nothing's convinced me that there is."

Nor did he budge four years later, after five of Florida's top climate scientists met with him to explain the science of climate change and why new policies were needed.

However, Scott reversed himself ever so slightly at the end of 2017, when he unveiled his $87.4 billion 2018–19 state budget. The governor proposed a $3.6 million outlay to the DEP to help local governments prepare for sea level rise problems—or 4/1000 of 1 percent of the entire budget.

By 2015 two other Florida GOP leaders, and contenders for the office of president of the United States, had turned their backs on the work of climatologists. According to the *Washington Post*, Marco Rubio once said he didn't "believe human activity is causing these dramatic changes to our climate." Jeb Bush, too, had by now become a "skeptic."

U.S. Democratic senator Bill Nelson was astounded by this sort of attitude among his peers. During a May 13, 2014, speech on the Senate floor, he implored all skeptical senators to stop being "politically correct" and recognize the truth of climate change.

Some high-level Republicans don't care what their party chiefs believe. In June 2014 four former administrators of the EPA who served under Republican presidents testified to a Senate panel that climate change was real. One of them, William Ruckelshaus, told lawmakers, "the four former EPA administrators sitting in front of you found we were convinced by the overwhelming verdict of scientists that the Earth was warming and that we humans were the only controllable contributor to this phenomenon."

Another speaker was Lee Thomas, who served under President Ronald Reagan. "We know that communities in our country are already dealing with the effects of the changing climate today," he said. "In my state of Florida we see increasing salt water intrusion infiltrating our drinking water supply due to sea level rise. Coastal communities are dealing with the impact sea level rise is having on their drainage systems, resulting in an investment of more than $300 million to upgrade flood mitigation infrastructure in Miami Beach alone. The economic impact is undeniable, and local governments struggle to address today's impacts of climate change while trying to anticipate the increased risk it poses in the future."

Most climate scientists are frustrated and angered by the delays and personal attacks aimed at them and their work. Of course, they understand better than laypeople that climate change is complex. They know well about the influences of the earth's orbital idiosyncrasies, sunspots, El Niño, and many other natural forces. Still, with almost complete unanimity around the world, they have concluded that the main forces of today's climate change are human caused.

And only humans can do something about those forces.

Floridians Take Action

Many Floridians aren't waiting for lawmakers to act. Across the state, various government officials, state bureaucrats, university professors, business leaders, environmentalists, religious people, and others are taking matters into their own hands.

The county commissions of Palm Beach, Broward, Miami-Dade, and Monroe Counties, for instance, have developed among themselves a "climate change compact" that provides a mutual blueprint for cooperative action and a way to seek federal funds to thwart the worst of climate change.

The City of Miami recently joined the compact. Like other Florida coastal communities, its leaders now have to figure out what to do. There are only three choices. Without world cooperation, stopping climate change isn't going to work. Even with that help it's probably too late to save a lot of Florida. Mitigating the harm from rising sea levels seems improbable, too, because Florida is so fragile.

That leaves adaptation, says Stuart Kennedy, director of program strategy and innovation for the Miami Foundation. "The impacts of sea level rise are already being felt here," he says. "Property is already being damaged. People's lives are being impacted. You can see it every day getting worse, especially during king tides [when the sun and the moon accompany high tide]. The days of flooding last longer."

Kennedy, a native of a four-thousand-foot-high West Texas desert, is pragmatic. He calmly explains that "we're not getting into the national conversation over whether climate change is real. Instead, we're moving ahead. We're interested in protecting the vibrancy of the community."

But how? There wasn't a plan in place, so the foundation's staff teamed up with counterparts at the City of Miami Beach and Miami–Dade County and applied as one entity to be included in a national program created by the Rockefeller Foundation that selects one hundred resilient cities around the world. Applying for the program was a three-year process. To be chosen, the South Florida group had to come up with a resiliency plan that took into consideration the views of the community and entities such as engineering firms, insurance companies, and Miami Beach officials that had experience raising roads and pumping seawater off streets. "The tricky thing," says Kennedy, "is that there is no example to look at. Yes, there's the Netherlands, but we're different from them, because of the limestone underneath Dade County; there is no way to build a dike."

Their hard work paid off. In October 2016 the Miami group was chosen.

But these South Floridians aren't the only ones taking action. Their allies include sea level rise "working groups" in Charlotte Harbor, the Indian River Lagoon, and the cities of Satellite Beach, Sarasota, and Tampa. All are working with the EPA's Climate Ready Estuaries Program to come up with adaptation programs for coastal areas.

In June 2017 Miami Beach hosted the annual U.S. Conference of Mayors. High on their agenda was the topic of climate change. Before they adjourned, mayors from around the country vowed to join and support international efforts to reduce carbon emissions.

Stealth Research

Though Florida's political class in Tallahassee is ignoring sea level rise, important climate research has been quietly under way for years at state universities and government agencies. For instance, the Florida Climate Institute—a consortium of the University of Florida and Florida State University—has formed a multidisciplinary network of national and international research organizations to provide scientific information about climate change that is Florida specific. The Florida Department of Economic Opportunity also has been publishing guides to help policy planners.

The University of Florida, Army Corps of Engineers, NOAA, and the Nature Conservancy have developed software tools to assist planners to come up with sustainable development practices. In addition, the Florida Fish and Wildlife Conservation Commission's climate change action plan is designed to safeguard Florida's wildlife.

Florida's water managers are also contending with vexing new questions associated with the probable disruptions of climate change. How will they conserve or recharge water in coastal-area aquifers contaminated by seawater? What new wastewater recycling problems will emerge? Will salt water intrusion outpace desalination's ability to keep up?

Another big question concerns the role of the private sector. Water-rich Florida had long been attractive to international companies that specialize in water utility management. Some observers argue that only corporate power—fortified with its economy-of-scale clout, state-of-the-art technology, and finely honed efficiencies—can keep Florida from becoming water-deprived. Others question, however, whether privately owned companies would remain in the state if sea rise makes engineering and water management tasks cost prohibitive.

Skeptics also doubt the private sector alone can produce a silver bullet for managing this most precious of natural resources. Questions of equity, quality, price, and environmental justice related to water must be addressed, too. That's why Florida needs a powerful consortium of civic, entrepreneurial, and political leaders to grapple with its water crisis.

Success will probably elude them, however, if they follow in the footsteps of those who've caused the problems. There might just be another way.

12

Rethinking Water

"TO A CARPENTER," goes the old saying, "every solution is a nail." A century ago, such thinking led men to believe they could engineer the water out of Florida.

Today, however, even the Corps of Engineers acknowledges the limitations of the carpenter's solution, as it notes on its website: "As a result of the engineering performed as early as the 1880s to make South Florida more inhabitable, the natural flow of water to, and through, the Everglades was severely altered. The construction of roads, canals and levees created barriers that now interrupt the natural flow of water that's necessary for the Everglades to survive."

Nonetheless, many modern water professionals and planners still believe today's water problems can be solved with engineering. They want to stay on what their industry calls the "hard path"—that is, the capital-intensive, supply-side approach to water management that relies upon water infrastructure designs, technologies, and decision-making rooted in the twentieth and even the nineteenth centuries. Think big dams and reservoirs, pipelines, pumps, gates, water treatment plants, and large-scaled, bureaucracy-laden public water utilities, districts, agencies, and international private water companies, and you've grasped the components of hard path.

Hard path wastewater systems also depend on "gray infrastructure," which Shawn Landry, director of the University of South Florida Water Institute, explains as the stormwater pipes and ponds that were put into place as a response to flooding, with the underlying attitude of "let's get the water away from us as soon as possible."

Need more water? The hard path response is simple, direct, and traditional: Go find new sources. Make the system more efficient. Drill deeper. Recycle. Desalinate. Drink toilet-to-tap products.

"But that's not enough," responds Peter Gleick, an environmental scientist, MacArthur "genius" fellow, and cofounder of the Pacific Institute, a nonprofit water research center. The hard path approach, he argues, "will not solve the problems that remain. We need new thinking. We need new approaches."

After all, water problems are different today. For the first time in history, human beings around the world are facing absolute scarcity and limits of their water supplies. New dams won't help much in drought-stricken California, where Gleick lives, because there's no surplus water to dam. More pumping of already stressed aquifers doesn't make much sense either.

There's also good news. Look around the world, Gleick insists, and you see people turning to—because they must—a new and more integrated approach to water challenges. And this is the way, Gleick and many others believe, to best deal with the world's water challenges.

He calls it the "soft path." There's nothing squishy about it, though. Those favoring the soft path concede that plenty hard path stuff—pipes, concrete, and so on—is still needed, but they want many hard path practices abandoned, modified, and complemented by a more enlightened, unified approach to water.

The soft path, like the hard, embraces supply-and-demand economics, but instead of simply looking for new supplies when demand rises, soft path water managers focus on their customers' water *needs.* To be sure, everyone needs water to drink, cook, bathe, and grow crops. But how much, if any, water do they really need for many other human activities? For instance, do they need *fresh* water to make more electricity for a never-ending array of energy-hungry electronic and electrical gadgets? Or to flush toilets? In fact, does anyone even need old-technology water-flush toilets at all, if clean and safe composting alternatives exist?

Do Floridians need to squander more than 50 percent of daily fresh water use on their lawns?

Soft path water consultants seek new conservation methods and technologies and advocate using smart appliances, tiered water-use pricing, drought-resistant landscaping measures, and innovative metering and monitoring devices, such as leak monitors and those that provide micro-irrigation.

They don't view wastewater and stormwater as liabilities. Instead, they consider them alternative water sources, along with rainwater harvesting and water farming—the practice of paying farmers and other landowners to capture water in reservoirs, as is becoming common in the Lake Okeechobee area.

Whenever possible, practitioners of the soft path implement new stormwater treatment systems that restore or preserve natural hydrologic functions. These natural systems keep rainfall in the local vicinity, allowing it to percolate into the local hydrology.

Imagine a hard path, or gray, stormwater infrastructure in a city. Rainwater pours off grimy sidewalks and streets into sewers and then is piped away.

Now visualize an alternative green infrastructure—one through which the stormwater passes into linked systems of eco-roofs, rain gardens, permeable pavements, public parks and recreation centers, urban wetlands, forests, and bio-swales, which are city landscape structures filled with vegetation and compost that remove silt and pollution from stormwater.

Soft path planners, in fact, strive to understand the hydrology of a construction site before the first shovel blade bites into the earth. How do the plants and trees at this site absorb moisture and release it? How much evaporation is at work? When is it at its peak? How is the water moving across and underneath the earth? Up, down, sideways, or not at all? What happens to the entire ecosystem when water is piped away as waste? How do biologic systems change at the receiving area? What if the mineral content of the piped water is different from the water it is replacing or supplementing?

They also take into account the fact that any community will change over the years—as will the local environment—until one day the hard path infrastructure for fresh, waste, and storm water is no longer appropriate. Soft path systems are set up to be adaptable to changing conditions.

Soft path advocates also argue that even if hard path engineers can supply water for everyone—and maybe even wring out a surplus by using new efficiencies—there's still the Jevons paradox to contend with. This effect occurs, say economists, when technological progress increases the efficiency of using a given resource—say, water. When that happens, demand may actually increase because the water is now cheaper and more plentiful.

Another way of looking at this issue comes from Chris Bird, Alachua County's Environmental Protection director. Consider, he suggests, a new subdivision with the latest in efficient toilets. They may reduce per capita wastewater usage, but if the number of residents also goes up, the total water withdrawal rises, too.

Many soft path techniques aren't new—just underemployed. But that may be changing. Many communities are shifting from gray to green, including Hamburg; Stockholm; Montgomery County, Maryland; Portland, Oregon; and Sarasota, Gainesville, and several other Florida cities.

Is there any pushback from engineers against the trend to green infrastructure? Yes, says Landry, but it isn't based on obstinacy to change. Nor does it stem from a generation gap among younger and older engineers. Instead, he says, "there is a reluctance to accept some of the new approaches because of limited data. People who traditionally treat water want numbers. For example, they might say this about a green infrastructure project: 'My computer model is messy with all these living things. Trees, for instance, have leaves and absorb water. But how many leaves will be on the tree next year? What if the tree dies? How do you measure the effectiveness of a rain garden?"

Another challenge for green infrastructure, explains Landry, "is that regulations that are meant for a system designed for water-bearing pipes don't always make sense with ones designed with organic, animated beings."

Thus, rethinking water regulations is evolving. But that's not all. Signs are strong that many people around the world are mulling over the importance of water, especially at the university level.

Water Goes to School

Many of Florida's academics, in fact, now spend a lot of their time rethinking and researching water. The University of Central Florida is home to the Stormwater Academy, and the University of Florida, the University of South Florida, and Florida International University have water institutes with a wide spectrum of academic disciplines providing an interdisciplinary approach to water issues.

"Water is moving way beyond engineering," says M. Jennison Kipp Searcy, a resource economist with the University of Florida Program for Resource Efficient Communities. "The approach is to break down the silos—the academic disciplines in which we work. For years, there's been a lot of talk about doing this, but it looks like now there's a real challenge for us to get to work."

Searcy once worked as a Peace Corps volunteer in Kenya, where people live close to the natural world. That experience convinced her that something vital has been lost in her own country when it comes to water. "We Americans are the outliers," she says. "We are so physically and economically distant from water. We don't see it; we don't understand it. There are systems underneath us. It's not just a big bowl under us. Much of the public can't see the connection between rainfall and land development patterns and how we respond to climate change. Our society is built on defending our individual rights, but we've lost that sense of community."

Attitudes may be changing, albeit slowly. As never before, communities and their leaders are realizing there are many stakeholders who have a right to have a say in water policies.

Not long ago, Searcy and other UF social scientists conducted a series of workshops that examined how agricultural producers, local government officials, and other water shareholders interacted over proposed water regulations for a river basin along the lower St. Johns River watershed. "We were trying to create an environment where everyone could feel free to talk," she says.

Sometimes the conversation at these sessions became heated, as participants argued about water allocations and computer modeling. Searcy noted some farmers became frustrated over costly mandates requiring them to reduce pollution. Environmentalists, on the other hand, often complained there wasn't enough regulation of agriculture.

Increasingly across the state, stakeholder meetings, such as this, are being held to give citizens the opportunity to at least be heard. These sessions should be more than an opportunity to vent, however. They should also inform stakeholders, while helping them find common ground and perhaps shape policy.

Because Searcy is an economist, she's always on the lookout for the lessons of "social marketing" and how policymakers can use these insights to nudge people in one direction or another in thinking about resource allocations.

Many utilities are studying these insights, too. A few in Florida now send their customers data that shows how their consumption of water and electricity compares with the average used by their neighborhood.

You might think consumers who used more water or electricity than the others would feel a wee bit ashamed and consume less next time. This does happen, but not always.

Studies, in fact, show that sometimes there's a boomerang effect, in which those people using less of a resource compared to their neighbors feel they are being shortchanged. So next month they'll use more to get their fair share.

Ideology may play a role, too. A 2013 National Academy of Sciences study, for instance, looked at how liberals and conservatives viewed purchasing energy-efficient compact fluorescent lightbulbs (CFLs). Participants from both groups told researchers they'd be willing to buy the bulbs if they were energy-efficient and saved money.

But the study also revealed the importance of choosing the right words. When asked if an appeal to "save the environment" would encourage them to buy the CFLs, liberals overwhelmingly responded yes. Saving the environment made them feel good.

Conservatives, on the other hand, were reluctant. It wasn't that they were uncaring about the environment; instead, they didn't want to be associated with what they suspected was on a "liberal agenda."

Water researchers hope that sociopsychological insights like these can assist policymakers in coming up with the right vocabulary when engaging with stakeholders in the future.

Looking at Values

Water's cultural aspects also intrigue Rebecca Zarger, a University of South Florida cultural anthropologist and a member of her university's Water Institute. Not too long ago, she participated in a tri-county interdisciplinary USF study that used questionnaires and personal interviews in Hillsborough, Pasco, and Pinellas Counties—the old Tampa Bay Water battlegrounds—to better understand the social institutions and personal relationships, views, attitudes, and behaviors at work in communities where water is transferred increasingly from a rural region to an urban one. Such insights could prove beneficial to water researchers everywhere, given that more than half of the world's population now lives in cities. This means more humans than ever before are using water from somewhere other than where they live.

"What we've found is that a lot of residents are very concerned about these issues," she says. "Those with higher levels of education are more aware of them, but others are also pretty well informed, too, because they've seen cypress trees come down and lakes dry up. Many people we spoke to felt passionately when they remembered swimming as kids in freshwater places that don't exist anymore. The longer they've lived in Florida, the more concerned they are. People are paying more attention to water issues than many water district managers and decision makers think."

However, when respondents to the USF study were asked if they had ever been involved in public action over water issues, a majority said no. Their lack of participation didn't mean they didn't care. Instead, it stemmed from a frustration and alienation from the status quo, Zarger explains. "In general, there's a feeling of being powerless and cynicism among the people we interviewed. They feel like developers and big companies always have their way. In general, however, there's also a big interest in having everyone learn to communicate better. There needs to be legitimate participation; that's when people contributing ideas feel as though they are taken seriously."

In the past, however, many stakeholders who have participated at water management district meetings stated they aren't taken seriously, and that the events are mainly perfunctory and merely allow citizens to blow off steam.

Jerry Phillips, director of Florida PEER, is cynical about calling everyone a stakeholder, because he believes it conveys legitimacy to polluters. That subtle change of vocabulary, he claims, began in the Chiles administration. "It was no longer viable to call violators 'violators,'" he recalls. "They were now 'stakeholders'; they had to be included in the conversation over pollution. So you had people who polluted now part of the discussion. That's when the seed [for trouble] was planted."

Nonetheless, Zarger believes water is fundamentally a social problem and favors hearing from everyone. "We know the science of water," she suggests, "but questions about transporting water from place to place, and how much environmental degradation people are willing to put up with are essentially social questions."

In early 2017 Zarger was asked to join a local Tampa Water Advisory Board. As far as she knows, she's the first ever social scientist to be asked to take part in policy decisions dealing with Florida water.

Low-Impact Man

One water expert who's been implementing the new way of thinking about water is Tom Singleton, a Monticello, Florida–based water consultant. Singleton has been on the soft path for many years, though he prefers to say he's a "low-impact" man. He's a forty-year veteran professional in the Florida water universe who has spent his career in the private sector and in the DEP and the South Florida WMD. As a state employee, he helped develop Florida's Total Maximum Daily Load program and assisted in writing the Florida Water Restoration Act. Singleton also took part in updating the SWIM plans, which outlined ways to improve at-risk water bodies in the Northwest and Suwannee River WMDs. He also was involved in the management of STORET—the state electronic water quality data archival system.

Singleton, it's safe to say, knows Florida's water regulations and a lot about water itself. Philosophy, biology, engineering, environmental ethics, design, economics, politics—all inform his comprehensive view of water. Such eclecticism is vital, he believes; otherwise, you won't ever really grasp the complexity of water issues. "We must think of water in a unified way," he insists. "As it is now, we put water into silos and manage it as if it were on farms. We tend to think of stormwater, wastewater, drinking water, and water for natural things as being different. But they are all interconnected. Water policy and management should reflect this approach."

Using an integrated approach, says Singleton, is to treat all aspects of water "holistically, as an interconnected system, and not at the expense of the other aspects." For example, "an integrated approach to providing flood protection might store floodwaters in historical wetlands for water supply, water quality treatment, and the protection of natural systems."

Singleton's ideas put him at odds with his hard path chums. "Too many engineers have lost a sense of interconnectedness with nature and, especially, the limits imposed by nature," he complains. "They have a certain hubris that they can build anything anywhere. It's a belief which leads to the McDonaldization of the world. The problem is that while in college, they were taught methods which they use on the job that increasingly do not apply. Their methods are based on the principle of statistical stationarity—the idea that past experience is a good predictor of future condition."

Disruption from climate change, population growth, pollution, and other forces may be undermining this principle. "Engineers cannot practice as they have in the past; new methods and a dynamic systems approach are required," Singleton insists. "Fortunately, it is a message that more and more engineers are starting to get."

Winter Haven Goes Soft Path

One engineer who does get Singleton's message is Mike Britt. He's the assistant utility services director of Winter Haven, a city of fifty lakes that sits over the Peace River watershed and the Floridan Aquifer basin. Eight years ago, Winter Haven had to come to grips with a big water problem. Its fifty-year-old water systems were antiquated and reflected the time and thinking in which they were conceived. In the past, there were abundant water supplies, so it seemed only natural to Winter Haven's former utility managers simply to expand the treatment and distribution networks to accommodate the growing number of people moving into the area.

City officials, however, realized those past assumptions no longer worked. They were confronted with a new reality of limits on the availability of existing sources of water. Winter Haven risked losing 4.7 million gallons per day (MGD) of existing permitted capacity, because of excessive pumping of the groundwater aquifer supplying drinking water. Furthermore, with additional population growth, experts projected Winter Haven would need an additional 2.0 MGD of potable water by 2035 and 6.0 MGD by 2050. That figure could go even higher.

"We'd known about these problems for thirty to forty years, but we couldn't really grasp what to do," Britt recalls. However, his job description meant he had to come up with an answer. He figured he had several assets to accomplish the task. For one thing, he was locally born and bred. "I knew water in my area," he says. "I knew where the drains were, where wetlands were, where to look for fossils. I also had an engineering degree, a master's degree, a scientific background, and first-hand experience. I also had a personal responsibility for the water supplies at the local level, and I knew I had to take a big look at what's going to work."

So he embarked on a two-year learning spree, consuming everything about water he could get his hands on. Finally, "I got to the point where I knew that the traditional approach wasn't the way to go. We had to change course and use a new and integrated approach to water."

That was when he contacted Tom Singleton for help. Soon, they came up with a water management and supply plan now being implemented in Winter Haven that has received national recognition.

"Avoid, minimize, and mitigate impacts to natural systems"—this is Singleton's mantra, and it set the tone and direction for creating a decidedly new water system for Winter Haven. The big goal was to both protect and enhance water sources to benefit both humans and the environment. For Singleton, this meant he had to help design a plan that restored the natural hydrology of the watershed. It also required reconnecting the region's lakes, creeks, wetlands, aquifers, and nature parks to provide a "natural infrastructure" instead of a traditionally engineered one that used reservoirs and pipes and such.

No longer, for instance, would the city get rid of "unwanted" water by discharging it far away. The idea is to store the water in the region's natural hydrologic systems for future needs and to control flooding.

Singleton's approach also takes into account the community's social and economic interests. So the city set up meetings where stakeholders—including representatives from government, business, environmental groups, and the general public—could meet and discuss their water-related concerns.

That didn't mean there weren't snags along the way. For instance,

Britt says, some landowners realized their properties were badly needed as components of a natural infrastructure and said they wanted to be "paid royally" by the local government for their land, or they'd sell out to developers.

This example shows what modern water planners are up against. It's not enough to know about hydrology, engineering, weather patterns, demographics, and such. As social scientists can attest, they must also deal with human beings.

Dealing with Human Nature

One of those who is mindful of human nature as he helps forge policy decisions in Tallahassee is Matt Caldwell, a state representative from Lee County. Caldwell is a rising figure in Florida politics with a penchant for water issues. As a young man, he worked as a volunteer for the local chapter of the Republican Party in his home county. "Around here the issue of the Everglades is ever present," he says.

After the summer of 2005, during which two hurricanes struck South Florida, local Republican Party leaders developed policy positions on water, especially those related to the Everglades and the Caloosahatchee River, and they chose Caldwell to explain them to the local community. This experience forced him to bone up on water issues and "raised my profile slightly, and in the process," he recalls, "I became a kind of water expert." Caldwell also worked as a volunteer on a resources board to help develop a Basin Management Action Plan (BMAP)—a kind of blueprint for restoring impaired waters—for the northern part of the Everglades area, requiring him to attend meetings from Key Largo to Ft. Myers.

Today as a state legislator, Caldwell has sponsored water-related legislation that has brought him both praise and criticism. All these experiences, he thinks, have taught him a lot about the politics of water. One of the biggest lessons could have come from social scientists. A person has to try to understand the values of others when political solutions are at stake, he says. There are cultural differences to keep in mind when applying the art of politics. For example, "South Florida has a real tradition of collaboration when it comes to fresh water issues," he says. "Everyone here is used to projects that deal with canals, levees, gates, and reservoirs. You couldn't live here without all the engineering. What we deal with in South Florida isn't over whether there's going to be a certain water project, but whether it's the most effective way of being successful."

But travel into North Florida, he suggests, and you will enter "a less populated, more rural, more multigenerational population with less experience in collaboration. It's a radical jump in terms of cultural acceptance. That doesn't mean you don't try, but when you don't have as many projects in your region, there's less experience. There's probably not a single reservoir in the entire Suwannee River valley."

Caldwell believes the most effective way to achieve any environmental goal is to help others achieve a cultural shift brought on by their own volition. "Religious conversion is longer lasting if it comes willingly rather than if it's been forced," he says. "Similarly, if you want some of these cultural shifts to be long-lasting and transformative, they have to take place in the same way."

Virtual Water

To really rethink water is to grasp a new and growing concept that was first introduced in 1993 by British geographer and water expert Tony Allan, recipient of the Stockholm World Water Prize in 2008, the Florence Monito Water Prize in 2013, and the Monaco Water Prize in 2013. This notion, at first, may strike you as strange. After all, you are asked to accept that some water isn't real. Economists call it virtual water. Many of them also say taking virtual water into account is mandatory for dealing with the challenges of worldwide water shortages.

It's hard to think about what you don't see. Groundwater, for example, or the fresh water coming into your home, or wastewater leaving it, travels in pipes buried or sealed within walls. You may not see it, but your power of reason lets you know it's there.

It's even tougher to conceive of invisible water that surrounds you wherever you go aboveground. But it's there. In fact, it's found in almost every object you come in contact with, whether an automobile, a pair of shoes, or a banana.

Think of oranges ripening in a Central Florida citrus grove. If they are harvested and exported out of the state, or even the country, the water contained inside them exits forever as well. It's as if the citrus water had been squeezed into bottles for a one-way trip. Over the years, millions of tons of citrus have ended up far away from Florida. This exportation of water, if it's not offset by an equal amount of water imports, becomes a water deficit for Florida.

If a foreign country imports those Florida oranges, it doesn't have to use its own local water supplies to produce them. This "saved" water volume can then be used for other purposes.

There's yet another way to think about virtual water. Compare a twelve-ounce glass of water with a twelve-ounce cup of coffee. Do both containers hold the same amount of water?

At a basic level, the answer is yes. But that's not the whole story, because the cup of coffee also includes virtual water. There were water costs associated with producing the coffee. Coffee plantations used lots of water to irrigate coffee bushes. Harvested coffee seeds needed washing. Packaging, transportation, and numerous other steps are needed to get coffee beans to market and eventually to make that cup of coffee—and each step has left a "water footprint." In fact, about thirty-seven gallons of water were required to make a typical cup of coffee. And that figure doesn't include the water needed to produce the cream and sugar.

In fact, virtual water was also used to produce the simple glass of water.

It was also involved in the production of the all-American hamburger—634 gallons, in fact. The water cost of your smartphone on the countertop? Nearly 3,300 gallons! The average American's direct water use (cooking, showering, and such) is about 100 gallons per person a day, but his or her water footprint is about 2,115 gallons daily.

Virtual water is increasingly becoming a factor in international trade strategies. Countries around the world, depending on the nature of their imports and exports, are gainers or losers of unseen water.

Some arid Middle East countries, such as Saudi Arabia and Israel, understand this water reality all too well. Economic policy in these countries and others now includes importing many water-embedded vegetables and fruits, rather than growing them domestically. It would be akin to millions of containers of bottled water entering the country to stay forever. These nations now have a net gain in water.

Northern European countries, however, also import more virtual water than they export, but they're not experiencing water scarcity; instead, they are protecting "their domestic water resources, land availability and land uses," according to the Water Footprint Network, a global network of businesses, government agencies, universities, and nonprofit organizations. "In Europe as a whole, 40% of the water footprint lies outside its borders."

As virtual water is becoming better understood, some nations are adopting policies aimed at limiting their water footprint. This means they face some big decisions. Should they, for instance, strive to be self-sufficient by producing certain goods inside their own borders? Or would it be better to import food containing virtual water and protect their domestic water sources for other purposes, such as electric energy production? What virtual water policies should be taken to help make sure all people on earth have a fair share of water in these times of scarcity?

Some water activists think Florida water supply planners, agriculturalists, business people, and policymakers should be thinking about such questions, too. The kinds of crops Florida farmers plant, for instance, shape water allocation needs. Is it good policy to grow highly water-dependent crops like corn in a state facing water shortages? Or sacrosanct Florida sugar? (It takes 475 gallons of water to produce two pounds of sugar.) What are the overall imbedded water costs when Florida adds to its manufacturing base? Increases its tourism base? Boosts electric production for new air-conditioned subdivisions? How can government and business sectors use virtual water insights to improve their supply chains and be more sustainable?

Cost of Water

The virtual water concept, of course, has its critics. Some point out that its proponents fail to recognize that water sources are not of equal value. There's a big gap, for instance, between expensively treated wastewater and purer rainfall. And what about geography's role? Just because you cut back on water allocation for a water-intensive crop in a fertile valley doesn't mean that the "saved" virtual water is easily made available in a distant desert for a different use.

Nonetheless, such questions should force Floridians to confront the actual market value of water. At the moment, many residents of the Sunshine State care so little for its water, in fact, that they practically give it away.

"No one pays a nickel for the privilege of extracting water from an aquifer, river, or lake in Florida," writes Tom Swihart, a thirty-year DEP administrator and author of *Florida's Water: A Fragile Resource in a Vulnerable State.* "After a water use permit is obtained, the only cost for withdrawal is a pump and the electricity to take the water out of the source. No payment has to be made to the state or water management district for withdrawing 10,000 or even 10 million gallons of water a day."

This holds true even for those who sell Florida water for a profit. More than sixty mineral or spring bottled-water companies, such as the Swiss-owned Nestle Waters North, pump millions of gallons a day from Florida water sources; they pay almost nothing. The companies then purify the water and sell it in plastic bottles, many of them outside the state.

Even the cost of their water permit is partially supported by state taxpayers. Florida's five water districts subsidize the cost of handling water permit applications. In 2011, according to Swihart, Suwannee River WMD recovered "only 16 percent of the cost of processing consumptive use permits."

Water is not traded in the market, and therefore "we treat it as if it has no value," says Jay O'Laughlin, a PhD policy scientist with the Indian River Lagoon Council. "But we're wasting it; we're throwing it away."

Florida receives 55 trillion gallons of rain a year and uses only 1 trillion gallons. And yet, says O'Laughlin, "We're already reaching limits."

This can be changed. Governments have the power to use taxes to encourage or discourage human behavior. They can also use water pricing in the same way.

How much to price water, however, raises a moral dilemma, according to author Alex Prud'homme. As he explained to *Rolling Stone* magazine, "Is water a common [resource], like the air we breathe? If it is, it should be free to everyone. Or, is it a commodity, like oil or gas that we process and sell in the marketplace? On the one hand, if you don't price water, people waste it. On the other hand, if we price it too high, then you are playing a game of life and death, predicated on making a profit."

Other ethical questions emerge when thinking about water. Few would deny that it is immoral, for example, to hand off to future generations depleted, polluted water supplies. Do humans also have a moral responsibility to keep all creatures and natural systems properly hydrated?

There's little doubt that engineers can produce safe, drinkable, recycled water. Some philosophers would say it makes ethical sense in a utilitarian way. "Recycling water means not only multiplying its uses, but also reducing the amount extracted from nature," writes Greenpeace water expert Klaus Lanz.

But many of Florida's water warriors wonder if future generations will be deprived of something precious if toilet-to-tap becomes the new drinking water norm?

The Ocala National Forest, for example, is loaded with trees, but it long ago ceased to be a naturally functioning forest. Instead, it has become, as intended, a huge tract of well-managed rows of commercial-grade pine trees. Similarly, a golf course is green and hilly and has ponds and plenty of open space, but who would dare to call it a glen?

Today, game management officials cull bear and alligator populations, not by the law of the wild, but by mathematical calculations determined in air-conditioned offices.

Is water, too, destined to become just another substance, like scrap metal or plastic, piling up for the next round of industrial cleansing and recycling? Will the most vital compound in the universe become yet another resource to be corporately managed to achieve maximum profit? Have Americans finally reached the moment dreaded by Rachel Carson, whose book *Silent Spring* launched the environmental movement? "In an age when man has forgotten his origins and is blind even to his most essential needs for survival," she wrote, "water along with other resources has become the victim of his indifference."

The coming water woes also force Floridians and people around the world to face an existential question. But it's not enough to ask if there will be plentiful water for future generations. The current generation must also take action to make sure that it's the kind of water worth having.

To help get the job done, they have only to look at what was done by Floridians who came before them.

13

All Is Not Lost

"AFTER ABOUT 10 YEARS [of living in Florida] you either give up and think about moving away," writes Malcolm Jones of the *Daily Beast,* "or you dig in for the long haul."

Jones lasted only thirteen years. "I have to say that I wasn't sorry to see Florida in the rearview mirror," he admits. "I also felt guilty, though, because leaving felt like giving up. But at the time I was so pessimistic that the notion of staying and fighting [to protect the environment] just seemed like an empty gesture."

It's easy to understand why Jones became depressed, if you've lived in Florida a few years and watched the Florida growth machine wipe out the paradise that attracted you in the first place.

Nonetheless, as many Florida veteran enviro-activists will tell you, citizens sometimes win. The jetport and the Cross-Florida Barge Canal victories are epic reminders of what can be done.

As gleeful antienvironmental and deregulation campaigns are now being unleashed at all levels of government, it's easy to forget that not long ago many Floridians did stand up for their natural resources and forged powerful tools to protect them. Just consider—in addition to the water-related legislative victories—the astonishing array of land acquisition programs created from 1963 to 2000: the Land Acquisition Trust Fund, the Land Conservation Fund, the Conservation and Recreation

Lands Program, Save Our Coast, Save Our Rivers, the Florida Communities Trusts, Florida Forever, and Preservation 2000.

Thanks to these people, more than six million acres of Florida are now protected for conservation or environmental protection purposes. Add federal lands and local conservation lands, and the total figure rises to about ten million acres of managed public lands, or about one-third of the state.

"Our programs have been successful for many reasons, the most important of which is the enthusiastic support, even demands, of our citizenry, who do not have to live in Florida for very long to notice treasured areas being lost to development at the alarming rate of 165,000 acres each year (an average of 453 acres daily) and who are keenly aware of the need to preserve our natural areas to provide a basis for our tourism-based economy," James A. Farr and O. Greg Brock of the Florida Division of State Lands wrote in 2006.

Some are federal-state partnerships like the South Florida Ecosystem Restoration (SFER) program, which includes the ongoing Comprehensive Everglades Restoration Plan and the Greater Everglades Upgrade. One phase of that complicated and massive undertaking—financed by the National Park Service, the Florida Department of Transportation, and the U.S. Department of Transportation—is the continuing effort to redo a section of the unintended dam known as the Tamiami Trail. Once completed, 6.5 miles within a 10.7-mile segment of the highway will be replaced with bridges. This breach will allow natural fresh water from the water conservation area north of the trail to flow into Everglades National Park south of the trail. The influx should help safeguard drinking water for South Floridians, protect habitat, and push back against saltwater intrusion. Bridges should also reduce road kills.

Meanwhile, the $1 billion Kissimmee River Restoration Project continues and is expected to be finished by 2019. By that time, a forty-mile-long winding river and twenty-five thousand acres of wetlands will have been restored. Despite setbacks caused by Hurricane Irma in 2017, results so far are promising. Dissolved oxygen levels in the Kissimmee waters, for instance, are higher than before restoration began. That's good news because higher dissolved oxygen levels help aquatic life to thrive. About 320 fish and wildlife species have returned to the river complex.

"So far, the Kissimmee River restoration is meeting or exceeding the South Florida WMD's research expectations. It is behaving as projected—with occasional surprises related to weather—such as hurricanes and high rain events that cause extremely high flow conditions," says Loisa Kerwin, director of the Riverwoods Field Lab, an education and research facility on the Kissimmee River, managed by Florida Atlantic University's Florida Center for Environmental Studies in partnership with the South Florida Water Management District. "The comparison of the C-38 canal with the restored river is very powerful."

Curious visitors from across the world come to the lab, including "teams of Japanese restoration hydrologists and Chinese wetlands ecologists," Kerwin says. "For the past twelve years, the United Nations Environment, Science, and Community Organization (UNESCO) has brought their scientists and managers to learn about the Kissimmee River restoration."

Other restoration projects are also under way. The South Florida WMD, for instance, is spending millions of dollars to reduce nutrient loading and enhance fresh water flow to the Everglades system from the Lake Okeechobee area to Florida Bay. It also aims to reestablish hydrologic conditions on thousands of acres of forest, swamps, marshes, and prairies.

Swiftmud, too, has designs for at least a dozen troubled water bodies within its basin area, including the Rainbow River, Charlotte Harbor, and Homosassa River.

The St. Johns River WMD, meanwhile, is reviving marshes and flood plains from the Lower St. Johns south to the Ocklawaha River and all the way to Lake Apopka, the headwaters for the Ocklawaha. The district says it is planning to increase flows in some areas, while reducing nitrogen and discharges into the many freshwater springs in its basin.

Springs protection also is on the agendas of the Suwanee and Northwest Florida WMDs. Many of the projects focus on advanced wastewater treatment, septic-to-sewer connections, and stormwater projects.

The state legislature is somewhat interested in springs, too. In 2016 it appropriated $50 million a year for twenty years for springs restoration and protection. According to the DEP, this brings the total for springs spending up to $135 million since 2013.

Springs supporters welcome the news, but they question some of the new projects. For example, some wonder why the state is upgrading wastewater treatment systems and getting rid of old septic tanks, while at the same time allowing local governments to permit the building of new homes with septic tanks.

Turnaround at Tampa Bay

One of the most celebrated Florida restoration projects is taking place in Tampa Bay. This water body spans almost four hundred square miles with a drainage area about six times larger. It opens to the Gulf of Mexico, nearly surrounded by six counties with a total population of 2.3 million people. Its estuary is shallow, only about twelve feet deep, but past dredging has produced a shipping channel forty miles long and forty-seven feet deep. More than one hundred tributaries flow into Tampa Bay.

In the 1950s Tampa Bay was pristine and productive. During the next two decades, however, it nearly crashed, as humans poured in by the hundreds of thousands. Pollution regulation was weak, so much so that untreated raw sewage and untreated stormwater poured into the bay. Making matters worse, developers ripped up mangroves to make islands for the building of residences and dredged the bottom of the bay to carve out shipping channels. Soon, vast, thick carpets of green algae did their dirty work along the shore and in the estuary itself, blocking off sunlight and eventually killing half the sea grass and harming the aquatic life that depended on it.

Tampa Bay was an environmental disaster, yet few seemed to care. In 1974 attitudes changed radically, however, after CBS's investigative news show *Sixty Minutes* exposed the estuary's problems to the nation.

Before long, embarrassed and motivated residents had mobilized and were pressuring their elected officials to do something. Their exertions paid off. The cleanup effort, in fact, became so big, successful, and impressive that Tampa Bay was included in the nation's estuary program. (In fact, the Tampa Bay Estuary is now home to some of America's most important bird nesting areas.) This distinction allowed the EPA to come to town and work with local governments and stakeholders to develop a comprehensive plan to address the big water quality and quantity problems.

Unlike the polarizing Tampa Bay water wars at the time, this massive cleanup effort was a major public and private partnership, nurtured by lots of good will. Heading up operations was a special forty-five-member committee, representing an array of government, business, and activist groups that brainstormed and came up with five hundred community projects to reduce nutrient loading to Tampa Bay. Utilities, for instance, modernized water treatment plants. New industrial discharge and fertilizer regulations appeared. Local power plants found ways to reduce the pollution they were emitting into the air, since atmospheric nitrous oxide contributed about one-third of the unwanted nitrogen into the bay. Phosphate companies learned how to reduce spillage of fertilizer they placed on ships at the Tampa dock.

A Canary in the Bay

All this work paid off. Today, a richness of sea grass—an indicator of estuary health, a sort of canary in the bay—has returned to the Tampa Bay Estuary. In fact, scientists recently counted more than forty thousand acres of sea grass, more than there was in the 1950s.

How did they do it? For one thing, the community worked together toward a common goal, says St. Petersburg environmental consultant Tony Janicki, who took part in the restoration. There was also vigilant regulation and monitoring.

How progress was measured was another reason. "You have to have numeric goals, so you know what you're shooting for, and not vague ones, which are so often used in restoration projects," Janicki says. The Tampa Bay team, for instance, had no use for inexact, aspirational goals, such as "the bay will show improvement."

Instead, it relied on numeric values, such as those used to count returning acres of sea grass. Having specific, concrete goals, Janicki points out, "makes it easier for decision makers to decide how money is to be spent."

Nobody is claiming that the work is finished. But all the same, thousands of Tampa Bay residents are proud of what they've done.

Tony Janicki is pleased, too. "Today Tampa Bay is always put up as the poster child for estuary programs around the country," he says. "But we can't just say we did our job, time to go home, things are great."

Janicki mentions this because Tampa Bay's population could double to seven million by 2050, and the last time the region experienced a population boom, the health of the estuary nearly collapsed. "It's scary," he admits, "but there are a lot of tools that have been put in place to address these problems."

Many of those same tools are being used to solve other human-made water problems elsewhere in Florida. In Sarasota, for example, many residents undertook a similar all-citizens-on-hand approach to reviving their own estuary. In 2009 their endeavors earned them first place in the EPA's Gulf Award program. Sarasota Bay Interlocal Partners and Citizens received praise for increasing sea grass levels by 29 percent over the 1959 level. At the same time, they achieved a 50 percent pollution reduction in the bay thanks to wastewater treatment improvements. Scallops also returned. The group also says that since 1989, it has restored more than sixteen hundred acres of intertidal and freshwater wetlands across the Sarasota Bay watershed.

Meanwhile, another successful environmental redo has been under way not far from the Magic Kingdom.

Artificial Wetland Superstar

Arrive early on a chilly, blue-skied January morning at the Orlando Wetlands Park, located south of Orlando in a rural area near Ft. Christmas, and you encounter a lush, swampy-looking landscape of cypress trees, cattails, and ponds. At first glance, it looks like the kind of place the dredge-and-fill boys would have loved to obliterate.

It also doesn't stink here. You might think that it would because this is where treated wastewater from the Orlando megalopolis bubbles up from an underground pipe to start its journey through what looks like authentic Florida marsh.

But this wetland park, like so much of nearby Disney and Company, is artificial. Not that anybody is complaining, however. Orlando is proud of its wetlands park, which has become an international showcase for demonstrating there's a managed and more natural way to clean up human waste than simply pumping the mess into a water body as Floridians did decades ago.

"We've had visitors from around the world," says Mark Sees, the tall, tanned biologist and director of the park, "including many scientists and engineers from Europe, Asia, South America, Africa, China, Ukraine, and Russia. Some come to learn how to handle some specific problems, like the Egyptians who are working with the Nile, and the Chinese with Lake Dianichi."

In 1984 Orlando city officials wanted to expand the Iron Bridge Wastewater Facility to keep up with population growth. The DEP, however, said that wasn't going to happen until the city solved its nitrogen and phosphorous problem. The city had already maxed out on the amount of these nutrients it could dump into the St. Johns River.

But with big growth on the way, "We had to look at a way to remove increasingly higher and higher amounts of nitrogen and phosphorous," says Sees. "The best solution was a natural treatment center. It was a controversial idea. At the time there were a lot of concerns about wetlands being used this way."

They went ahead with the idea anyhow and built the world's first large-scale artificial wetlands to treat urban wastewater. First, they bought fourteen hundred acres of cow pasture. Workers contoured the land and constructed miles of earthen dams, berms, and traps to keep water in. They hand-planted more than two million plants and twenty thousand trees to "polish" treated water sent over from the Iron Bridge facility. The vegetation does this by absorbing leftover amounts of nitrogen, phosphorus, and other contaminants from the thirty-five million gallons that flow through the wetland park every day. At the end of the process, the water enters the St. Johns, cleaner, say wetland officials, than the river itself.

Today, the wetland has also become home to more than two hundred bird species, fish, alligators, snakes, turtles, and predators such as raccoons, bobcats, otters, deer, black bear, and even an occasional panther. "We provide habitat for threatened and endangered species on both federal and state lists," says Sees.

It's a recreation center, too. Arrive on a weekend, and you can ride bikes, jog, or hike around the park's twenty miles of roads and woodland trails that pass through marshes and hardwood hammocks and along scenic lakes. On Fridays and Saturdays, you can also take a guided tour on an open-air tram.

The Orlando Wetlands Park may be the biggest of its kind in Florida, but similar undertakings, though usually much smaller-scaled, also have appeared in Alachua, Polk, Liberty, and Levy Counties and in the cities of Delray Beach, Gainesville, Pembroke Pines, and Ocala.

In addition, the South Florida Water Management District has built Stormwater Treatment Areas (STAs), which are man-made wetlands whose vegetation can absorb up to 80 percent of the phosphorus from runoff of nearby agricultural areas before heading south to cleaner waters in the Everglades.

A One-Man Show

Entrepreneurs are also at work too, doing their bit to restore Florida and make a profit. Terry Bastian, for instance, is a Jacksonville-bred builder and artist who started his "ecological design" career building Zen gardens before graduating to restoring ponds and wetlands in the greater Boston area.

Years ago, after an accidental exposure to floor sealant fumes poisoned him, he moved about aimlessly, suffering from a kind of dementia, he says, until he ended up working at the Cotton Tree Lodge in the jungles of Belize. There, he met a Mayan shaman who gave him a drink brewed from a special vine, which Bastian says cured his mind.

In 2014 Bastian moved to Dania, Florida, to get a fresh reset on life. One day he learned about a pest control company that was using bacteria to clean up empty poison tanks. Because of his own bout with living in a "toxic universe," the company's approach resonated with him in a personal way.

Bastian wondered: Why not use the same technique to clean up the toxins afflicting America's waters? Today, his company, Slurpits, produces sea-serpent-looking biofilters covered with a synthetic material that absorbs oil and pesticides, but rejects water, which makes the serpent float. "Inside," says Bastian, "is a blend of selected non-GMO, EPA-registered bacteria, archaea that break down hydrocarbons and other toxins, mimicking the way nature works."

Instead of shipping production overseas, Bastian has teamed up with the Palm Beach Habilitation Center to employ disabled Americans. His serpents have been appearing in golf course lakes, storm drains, ditches, and other waterways. Two of his more notable undertakings are an earth serpent that protects the water supply of the city of Cambridge, Massachusetts, and two giant sea serpents in Salem's harbor.

Taking It to the Roofs

Many Florida roofs are now sucking up great amounts of nitrogen and phosphorus, thanks to Bold and Gold (B&G), a pollution-control product developed by the Stormwater Management Academy at the University of Central Florida and based on the research of Marty Wanielista, professor emeritus at the academy.

B&G is a contrived soil made up of special minerals, organics, and clay. Lemongrass is planted to hold the soil in place on a waterproof membrane that is fitted to a rooftop.

Wanielista claims that up to 70 percent of nutrients is absorbed from rainwater that soaks through the pollution-absorbing roof turf. Today, green roofs adorn UCF's campus and many other buildings across Florida, including a thirty-three-thousand-square-foot application at the One-Stop Permit building in Escambia County.

The patented soil is also being used as a filter inside metal boxes being put into municipal stormwater ponds, infiltration basins, swales, containment tanks, and drain fields across the state.

So far, the biggest application of UCF's special soil is a forty-three-thousand-square-foot tract of land in the City of DeLand, where it is used as a final step for partially treated sewage.

Use of the product, says Wanielista, is spreading across the country, and the academy is receiving inquiries from as far away as Germany and China.

B&G and Terry Bastian's serpents are but two of the many innovative new approaches that keep appearing to help Floridians clean up the nutrient mess. Some of the other approaches utilize artificial "floating wetlands" that use algae itself to absorb nutrients and thus help to clean up.

Mechanical methods of algae removal are also being used. In Ft. Myers, for instance, workers on a boat deploy a skimmer to vacuum up algae from the St. Lucie River. Next, a truck hauls the gunky mess to a water treatment facility where it is cleaned and deep-injected into the earth.

Meanwhile in Sarasota, Mote Marine Laboratory, an independent research institution, is testing a process that injects ozone into red-tide-spoiled water to break down the toxins.

Looking for a Sunny Side

Along with the examples of the past and actions being taken today, are yet other reasons to be guardedly optimistic about Florida's water future. For one thing, the Growth Management Plan is still on the books. All that the Scott administration did was to kill the implementation of it, not the law itself. A bounty of powerful land acquisition and water protection laws are also still around.

In 2014, 75 percent of Florida voters passed the Florida Water and Land Conservation Amendment, which authorizes the state government to spend $300 million to purchase environmentally sensitive lands, including "lands that protect water resources and drinking water sources."

Thus, the measure is now embedded in the Florida state constitution; however, that doesn't mean the will of the people was respected by the Florida legislature. For two years in a row following passage of the amendment, a majority of state lawmakers brazenly used the funds provided by the new amendment for other purposes than land acquisition.

However, a coalition of environmental groups went to court to stop the legislature. On June 15, 2018, a Leon County judge ruled in their favor and ordered lawmakers to spend the money correctly. Some lawmakers vowed to appeal the decision, but for the moment it seemed the battle to protect Florida's fresh water was headed in the right direction.

David Guest, the Earthjustice attorney, has spent much of his legal career fighting water polluters in court. Yes, he concedes, Florida's environmentalists have experienced big losses and setbacks in recent years. Nonetheless, Guest is confident about the future and expects things to get better, but not all at once. "Time marches progressively," he says. "You don't hit a home run every time, but you also never get to first base if you don't swing."

Some observers even think Florida's water management districts may be slightly changing for the better. Despite their controversial decisions, along with the budget cuts, the personnel purges, and the allegations of political interference, "the state's approach to managing water has evolved significantly," says Tom Singleton. "What started out in the '70s as a top-down, command-and-control approach to regulating specific pollutants and problems has given way today to a more collaborative, holistic approach for dealing with the cumulative effects of land uses in watersheds, the restoration of natural systems, and the purchase of environmentally sensitive lands. Despite substantial progress, impacts are still occurring, and much work remains."

There's still much work to do. And the lion's share of that labor is needed in that most treacherous of human arenas: the world of politics.

14

Politics and Solutions

IT'S OFTEN SAID that you can't see the solution unless you first see the problem. If so, Floridians are making progress, because a significant and growing number of them do recognize their water problems. How else can you explain that three out of four voters—4.2 million strong—approved the Water and Land Conservation Amendment?

Thus, it's now time to see the solutions—and implement them. They aren't hard to find. For starters, the water management districts have provided Floridians with their water pie portfolio offering a range of hydro-actions.

Conservation is one of the key ingredients. Among other steps to get this done, Floridians must abandon their current water-guzzling habits.

New reservoirs above and below ground also must be created, of course, but the state also needs to protect its wetlands and buy more conservation lands to capture, store, and cleanse water; letting nature do these things is less expensive than achieving them the hard-path way.

Taxes and bonds aren't the only way to pay for these acquisitions. Some water experts suggest that public utilities also could raise money with market offerings, such as lifetime annuities.

Preserving and reestablishing the state's hydrology may be the most important step Florida must take to assure a water abundance. As Tom Singleton advises, "all open spaces, and especially within urban areas, must be reconfigured, and all new development must be designed to capture, store, and cleanse—to the maximum extent possible for each site—up to 100 percent of the rain that falls on it."

Floridians ought to know by now they must also clean up nutrients and contaminants from their water supplies. "The state needs to set mandatory, enforceable targets on water quality for farmers and other industries," advises the *Tampa Bay Times* editorial board. "It should also press for an end to American price supports for sugar, which only subsidizes an industry that has contributed significantly to the environmental degradation of the basin."

Individuals can do their part by no longer fertilizing or watering their lawns. Larger-scale problems, however, require larger-scale solutions; this means implementing stricter compliance and enforcement regulations for not only agriculture, development, industry, and commerce, but everyone. Annual inspections of all septic tanks should be obligatory.

Is it also time to look at the fairness and usefulness of current water allocations? Many water activists think so, and call for that analysis to be guided by the lessons of virtual water. For example, with huge water scarcities on the way, shouldn't water gatekeepers reevaluate how much water flows to the Florida sugar industry, which in 2005 used 875 million gallons a day, or one-eighth of all the water used in the state?

Many water policy analysts, meanwhile, think it's also high time to establish a universal, statewide tiered pricing system for water in order to encourage conservation, though some, such as Heath Davis, mayor of Cedar Key and a former Suwannee River WMD governing board member, worry that doing so could reignite the water transfer controversy. Thirsty South Florida, he says, might find it a better deal to purchase North Florida water and transport it than to develop costlier local water sources.

However, Tom Swihart, former head of DEP's Office of Water Policy, makes a strong case for enacting fees. Miners have to pay an excise tax to Florida for extracting minerals, he points out. So why not make people pay for water withdrawals? A fee for "different regions or hydrologic circumstances" would allow water management districts, for example, to charge a "drought surcharge" when rainfall is slack.

And why not add an array of varying fees for different classes of water use? He suggests not just classes such as commercial, residential, and agricultural, but also those that include, for instance, the time of day or distance required for the transportation of water.

"A fee of a few nickels per thousand gallons would not be a major increase in the cost of water," Swihart writes, "particularly if there were a compensating decrease in property taxes." Having to pay would also remind Floridians that water comes with a cost and would in theory persuade them to conserve.

Florida residents already are accustomed to hearing their water management districts tell them to conserve water by washing their cars only on designated days and so forth.

Such admonitions, however, rankle many homeowners because they are fully aware that at the same time these districts are urging restrictions on homeowners, they are busily issuing brand new consumptive use permits like library cards to big developers, even when springs, rivers, and lakes are receding.

The Florida growth machine may be happy about this apparent inconsistency, but many others aren't, as massive public support for Hometown Democracy not long ago indicated. As a result, there are also increasing calls for imposing something that's an anathema to the Florida growth machine—that is, a moratorium on the issuance of consumptive water permits.

As one advocate of the idea put it: Why can we put a limit on the number of bears in Florida but not a limit on water permits?

In May 2011 former state senator Evelyn Lynn, a Daytona Beach Republican, did more than wonder. She introduced a bill—perhaps the first of its kind in the country—that would have imposed a moratorium on commercial water extractions statewide for a decade. In 2012 it died in committee. Nonetheless, the fact that a Florida legislator had actually championed such a moratorium cheered many water activists, who hope other lawmakers will try again.

Broken Politics?

Making any changes to Florida's water habits and practices—perhaps even imbedding them in a revamping of the Water Resource Act of 1972—requires legislation, hard work, funding, and commitment.

"But right now we don't have the leadership to go after these problems," complains Lisa Rinaman, the St. Johns Riverkeeper. "In this current political environment there's a lot of demonizing going on in Tallahassee. Years ago, we would sit around and think together to solve a problem; but now we've lost the sense of how we can focus on the greater good in both the economy and the environment. I've been trying to wrap my head around the idea of how can we make a difference. At least we should be able to agree on water conservation with reasonable regulation that protects our water system. Our environment should not be a partisan issue."

The rancor, however, isn't simply between Republicans and Democrats. "I think one of the greatest threats to water in Florida in recent years," says author Cynthia Barnett, "is the intense political acrimony over every single water issue. I find it most painful that Florida's agricultural and environmental communities cannot seem to find common ground."

Pensacola's Laurie Murphy, the executive director of the Emerald Coastkeeper, believes water problems where she lives won't be solved unless the local political system improves. "A big problem in the Panhandle is that we don't have trust in government," she says. "We don't cooperate, and we don't have anyone willing to take a risk to try something different to solve some of our problems."

Distrust in government isn't limited to the Florida Panhandle, of course. It's ubiquitous. Ironically, during the radical 1960s, it was the political left that railed against the power of Big Brother. Today, that antigovernment sentiment has been appropriated by the political right, which rants against the nanny state twenty-four hours a day.

But there are also many Floridians who believe the private sector now exerts too much control over government at all levels. Others believe just the opposite. Fueling cynicism in both camps are the lies and distortions arising from the dark world of fake news and propaganda that digitally zip through the phones and computers of the world.

Thus, more than ever, Floridians need water policies based on facts they can trust.

You've Got Your Scientists, and I've Got Mine

The go-to people for those facts include a wide range of Florida's scientists—in academia, government, and the private sector. They're the ones who provide the technical studies and reports used by the WMDs, DEP, and other regulatory agencies that determine the fate of Florida's fresh water.

When not in the field, they can be found indoors, crunching data, running computer models, teaching, attending conferences, speaking at public forums, keeping up with new research. They publish the results of their own research and then steel themselves for the chastening commentary and painful criticism of their peers, foes, allies, and the public. That's just the nature of science. Any good scientist will tell you that it's a never-ending process.

However, if you talk to enough of these highly credentialed men and women, you're apt to hear sounds of discord that go beyond what usually takes place within the profession.

For one thing many of them are offended and angry. Hundreds formerly worked in state agencies and now complain publicly and confidentially that they or their colleagues were fired, demoted, or slandered, not because their research was wrong, but rather because it contradicted the expectations of powerful people.

The fear of such treatment reportedly still lurks in the minds of many currently employed scientists and state employees charged with protecting Florida's natural resources.

"After so many years, I've lost trust in elected and high-level appointed officials, especially the governor," laments former springs biologist Jim Stevenson. "The purpose of the government today seems to downplay the seriousness of the problems" facing our springs and waterbodies. "Self-preservation is what's on the minds of state employees. They don't dare speak out."

Gary Goforth, the Everglades environmental engineer, believes he was punished for speaking out. "I've worked for the state of Florida over twenty years, and did a lot of work for the water management districts," he recalls. In 2013 he appeared as a private citizen at a water management district public hearing and spoke against a district policy decision. Not long afterward, he claims, "$50,000 worth of work of a portion of one of my contracts was cut out."

All the conflicts and disagreements over Florida's water woes, he adds, "are fueled by politics and greed, which are often the same thing. Great scientists like Bob Knight remain woefully out-funded in their fight to educate the public and policy makers."

Alleged political interference in the work of science isn't the only complaint you hear when you talk to Florida's water scientists. There is also a reportedly high level of tension among the scientists themselves. Some are said to have accused some of their colleagues of incompetence, or becoming sellouts, meaning they skew their findings to please political or corporate power.

Some staff scientists within the districts and the DEP bristle at outside charges they are using flawed science to make Florida's water bodies appear healthier than they are. At the same time, some of these same staffers are contemptuous of independent scientists who speak out against what they see as bad water policy. All too often, these critics stand accused of being advocates for various causes and having lost their scientific objectivity and credibility.

The atmosphere among scientists who work on water issues in South Florida can be especially fraught with tension, says FAU's Brian Lapointe. He recalls one tumultuous meeting in the 1990s, where his colleagues squared off verbally over whether it was a good idea to pump fresh, but nitrogen-rich, water into Florida Bay. Would a reduction in salinity be beneficial enough to offset an influx of dangerous nutrients that were likely to make things worse for the bay?

The question evoked vitriol, he recalls: "The ideological divide at the time was as bad as a Trump and Hillary debate. A Canadian scientist who was present later told me he refused to come here anymore and advised me to get the hell out of Florida."

Lapointe, of course, remained in Florida and today is deeply involved in Indian River Lagoon research. But he says the tension is still ever present. Honest differences and ideology aren't the only things that drive a wedge between scientists. "Money has also changed everything," Lapointe says. "With funding cuts, many scientists are quick to grab what they can even with strings attached." As a result, he adds, a lot of time and money are wasted on projects that were dubious to begin with.

At times, the problem isn't what's said, but what is not. "The people at the water management districts and the DEP are in a quiet mode. They're keeping their heads down, until the political tide changes," says Jan Allyn, a content manager who collects water data from across the state for the University of South Florida's Water Institute. "They are cautious about giving their opinions, but the sad thing is they are also cautious about saying what the facts are."

Some Florida scientists say any divide among them is less about differences on issues of science and more about the policy implications of scientific findings. They point out that water management districts, which once operated with independence, have become hostages to political partisans who put their own political and economic agendas ahead of environmental protection.

In contrast, they say, look at the United States Geological Survey. It functions without interference and produces scientific data used and trusted by government agencies at all levels.

All this fuss and rancor frustrate those trying to fix Florida's water woes, because what they really want is information they can trust to make sound decisions. Of all the ways humans have of knowing about the world, the scientific approach has emerged as the best. It has trumped competing methods of inquiry based on authority, experience, faith, and even logic.

Science, of course, is a human construct, and therefore imperfect. It is always subject to challenge and peer review. This process culls out the junk and green-lights further inquiry into what's left standing. Scientists like to say they rarely disagree on basic data; it's theories that give them trouble.

However, it's only normal that Florida's water researchers disagree about what data to use; often this is because they use different modeling and methodologies. And it may also be true that in some cases politics, money, and fear affect the judgments of scientists and policymakers. If so, how is everyone else supposed to have confidence in their recommendations for coping with water issues? After all, isn't that why university scientists have tenure? To make sure they will be immune from outside pressure? Why, then, shouldn't scientists at Florida's water management districts and the DEP enjoy the same protection?

A State Academy of Sciences?

One possible remedy is that Florida should establish a state version of the National Academy of Sciences that would serve as a kind of science clearinghouse.

It's an idea that has precedents, but not encouraging results. In 2016, for instance, the National Academies produced an Everglades' progress report whose message, according to Gary Goforth, "is clear and unequivocal—more lake water needs to be sent south. End of debate."

Of course, that debate is far from over.

In the early 1990s Brian Lapointe, with help from the South Florida WMD, helped organize a panel of eminent national scientists to review the science related to the health of waters and coral reefs in the Florida Keys. However, he says, it didn't really help much because politics trumped science. "This same thread can be traced to the current state of denial about sewage pollution and pointing fingers at agriculture, especially Big Sugar," he adds. "It's actually not about the science or the environment; it's all about the dollars."

Janet Llewellyn, the former DEP administrator, however, doesn't think a new statewide body is needed. "In my experience," she says, "the WMDs do frequently provide for independent peer reviews of their models and things like their minimum flow and levels technical work—especially if it is expected to be controversial."

What Jim Gross, the former St. Johns River WMD hydrologist, suggests is refinement of that process. He envisions "a better institutional approach to independent peer review of the scientific work by resource management agencies like the water management districts." He's not talking about expert witness work for administrative hearings. Instead, "I'm contemplating ongoing, careful, thorough, and deliberative review of data, methods, and models. The people doing the review cannot have a financial interest in the outcome of the reviews. All that can be on the line, so to speak, is their professional integrity. So, yes, something similar to what exists with our prestigious National Academies would be a good approach for Florida."

Recover What's Been Lost

All of this concern about getting the facts right explains why the districts need to undergo changes, says Saundra Gray. A former owner of Gemini Springs Ranch, she was appointed to the St. Johns River Water Management District governing board in September 1985 by Democratic governor Bob Graham and reappointed by Republican governor Bob Martinez.

"Things in those days were not so political," she recalls. "People were more civil. We had a more balanced board, too—more environmentalists, like people from the Sierra Club and Audubon. The governing boards don't have people like them anymore. The boards need to diversify. You must have different types of people to make good decisions. And most importantly you need people who are educated in water. We do know the answers, but it takes political will and a good leader."

A return to proper funding is also needed. It's not hard to believe that Florida's water management districts suffer from the budget cuts and firings made not long ago. Audubon Florida is calling on the state to address this issue by restoring the "$200 million a year that has been cut from" district budgets.

Another restoration that can't be ignored is reinstituting strong, state-run growth management.

Tom Swihart recommends bringing back the nine-member Florida Energy and Climate Commission, destroyed by Governor Scott and the state legislature in May 2011. The commission's job was to put into action the state's climate change mitigation and adaptation programs. Swihart also thinks strategies for dealing with climate change should be included in the state water plan.

Many reformers also want greater transparency at the water management districts and more governing board meetings held at locations other than at agency headquarters. They also advocate that executive directors be required to have strong management skills and backgrounds in water management and that hiring of staff should be based on knowledge, skills, and abilities, not on political cronyism. They also hope for a comeback of credibility and public trust in the districts.

"Here's what I've got a problem with: Paying taxes to a district controlled by industry cronies who prioritize special interests over everyday people," writes Eve Samples, an opinion writer for the Treasure Coast Newspapers organization.

One way to go after this stacked-deck system, suggests Samples, is to stop governors from making these selections. Instead, let voters elect board members. After all, she reasons, shouldn't voters have power over any agency with taxing authority?

Any reform, or restoration, of course, must first be legitimized in Tallahassee, a city teeming with powerful corporate lobbyists, many of whom consider Florida's water sources as their personal resource base.

People Power

The obvious solution to any political problem, of course, is to elect honorable, capable, educated, smart, tough people who fully understand environmental issues. In a state where money is power, that can be hard to do.

If Florida's lawmakers can't, or won't, fix problems, citizens can grab a very big hammer that's always at their disposal when the legislature ignores them—the citizen initiative process. The Florida constitution grants them the right to use the petition process to put constitutional amendments on the ballot. That's how Floridians were able to propose and ratify the Water and Land Legacy Amendment.

The state constitution also requires that a constitution revision commission meet once every twenty years. The commission's task is to review the constitution and place amendment proposals on the ballot for voters.

When the 2017–2018 Constitution Revision Commission convened, many Floridians were pleased when one of the commissioners, Jacqui Thurlow-Lippisch, proposed an amendment guaranteeing Floridians the "Right to a Clean and Healthful Environment." The idea came from students and professors at Stetson and Barry Universities.

The proposal, explains Thurlow-Lippisch, a former mayor of Sewall's Point in Martin County, would amend Article II, Section 7, of the constitution "to add that every Floridian has the right to a clean and healthful environment and that any person may enforce this right against any party, public or private."

All too often in Florida, when citizens become too frisky, an array of business and industry groups are quick to push back. This time their ranks included wealthy lobbying heavyweights, the Associated Industries of Florida and the Florida Chamber of Commerce, who expressed horror and hired expensive attorneys to fight the proposed amendment.

"They said they were afraid of an onslaught of citizen lawsuits," says Thurlow-Lippisch, "but the truth" is the business community got the upper hand during the recession, when jobs were so important, and now "they're on a roll, and they don't want anything to get in the way of what they're doing."

Those business groups got what they wanted this time too; on January 12, 2018, the Judicial Committee of the Florida Constitution Revision Commission unanimously rejected the amendment.

But the issue isn't dead. It can be revived as a citizens' initiative. If it were ever ratified, Floridians would have sturdier legal ground to stand on to object to the defilement of their water resources.

To be effective, however, citizens need to know the issues. For many that means getting a big dose of water education.

Getting Armed with Water Lore

To obtain that information, somebody first has to be watching what's happening to Florida's water bodies. "We don't know exactly how our water situation is getting worse, or if it's getting better, because we're not monitoring these very important indicators anymore," says Lisa Rinaman. "There's been a severe reduction in monitoring throughout Florida, some of it due to budget cuts, and some of it, I think, is because some people don't want to know if there's a problem. That's something we need to resolve. We can't address the problems if we're not monitoring, and not understanding the conditions of these waterways."

Miami Waterkeeper Rachel Silverstein knows well that absence of monitoring. In November 2017 it was up to her office—a mostly one-woman operation—to publicly expose that authorities had ignored a leaky sewage pipe dumping filthy waste into Biscayne Bay for more than a year.

Information, even if it is insufficient, can't remain in academia or in the hard drives of water management districts. All Floridians have a right to know about their water. One way to learn would be to attend well-advertised, state-funded, free, local and regional water forums whose message is delivered in lay language for members of the general public. At such events—held when working people could attend—top scientists, engineers, water experts, journalists, academics, water planners, and activists could try to educate and update their fellow citizens and elected representatives about the state's water problems and solutions.

Here, too, citizens could ask basic hard questions about water politics. Who, for instance, really controls Florida's water supplies? Is the state cracking down enough on polluters? When will the state legislature get serious about sea level rise? Why do so many of Florida's leaders advocate relentless growth even if it means having to deplete vulnerable water supplies?

Citizens could also learn what their own responsibilities are as water stewards for future generations and the natural world.

Learning something new often breeds astonishment and sometimes anger. And that anger can lead to change. In fact, an aroused, if not outraged, citizenry sometimes demands and gets change, even if it costs money. The algae blooms, for instance, caused such an outcry in Brevard County that the "county approved a half-cent sales tax for 30 years to generate $300 million to clean up the lagoon and stringent regulation governing septic tanks and fertilizer," observes Florida economist Henry "Hank" Fishkind. "And it's pretty remarkable for a conservative county to act on its own to address real dangers to its citizens and its economy and to be willing to spend $300 million of its own money to do so."

Reform, as the jetport battle proves, sometimes flows from the bottom up. All too often, however, it's tough for grassroots groups to fight for water justice with just a few dedicated people and a cigar box half-filled with nickel-and-dime donations, especially when you face huge corporations and water districts with an array of attorneys. That's why many Floridians seek out established, seasoned environmental organizations.

Florida has scores of them. Much of the heavy lifting—legal challenges, research, and fund-raising—is done by well-known organizations such as the Sierra Club, Audubon, Florida Defenders of the Environment, the Florida Wildlife Federation, the Everglades Foundation, the Conservation Trust for Florida, and Earthjustice.

There are yet many others. A sampling: 1000 Friends of Florida, the Florida Conservation Coalition, the Riverkeepers, Rainbow River Conservation, the Miami River Commission, the Indian River Land Trust, the Northwest Florida Environmental Conservancy, and the Howard T. Odum Florida Springs Institute.

All for One and One for All

Perhaps the most potent political action any Floridian can take is to help elect a governor, because that individual wields tremendous power over Florida's water resources. The governor has veto power over legislation that polluters and water wastrels want to pass. He or she appoints the secretary of the DEP, all the governing board members of the five water districts, all the judges who preside over administrative hearings where challenges to pollution laws take place, and all the members of the Environmental Regulations Commission.

"It's how the governor thinks that dictates how these agencies behave, whether or not they let a permit through, whose permits they allow through, whether or not they're going to enforce pollution control laws in a specific area or water body," says environmental attorney Heather Obara. "It all depends on how the governor is being influenced in Tallahassee."

Individual action is essential, but in most cases there's strength in numbers. In the nation's third most populous state—a widely diverse one—Floridians have no choice but to work together to solve their water and environmental problems.

State Representative Matt Caldwell agrees with this premise. However, having seen too much strife in the state over Florida's environmental problems, he believes that solving them can be achieved only if comity, not enmity, is in play.

"At the end of the day, it's not going to be accomplished by being divisive and fighting each other within the state," he says. "We have to be a unified front to achieve these policy goals, and we're not going to be able to wait for someone else to show up and fix those things. We're going to have to be the ones to make things better."

But what will it take to motivate everyone to take action? "We're a crisis species," says springs expert Jim Stevenson. "We wait until the last buffalo is about to go off the cliff before we do something."

If so, the pressure to act may come only when enough Floridians, like the residents of Cedar Key, wake up one morning and find their fresh water supply is gone.

When they do, they may want to learn all they can about what occurred at the fishing community on the gulf.

"Three things happened that day," recalls Cedar Key's mayor, Heath Davis. "There had been a three-day Memorial Day celebration, which meant the water supply was under pressure from a lot of out-of-towners. There was also a high tide that day that brought in salt water. And a drought was under way, which had already caused low water levels."

These three troublesome vectors—increased demand for water, sea level rise, and drought—are also converging in many other Florida communities today. That may be why Davis is being asked to speak around the state about his Gulf Coast fishing community's water experiences.

Cedar Key hasn't dodged a bullet, but residents are learning to cope. And they aren't denying the threats to their fresh water supply, the physical structure of their small city, and even their way of life.

A leader, says Davis, has to have good conversations and trust with constituents: "Do that, and you can help them navigate through these strange times."

There's one more thing, he adds, that makes Cedar Key different from so many other Florida communities. It may be the small city's saving grace. "There's not much money here in Cedar Key," he explains. "There's no financial benefit that can go to anyone because of our water problems. That's why we're able to work together."

Davis's message may be the most important one Floridians and their leaders can ever hear—that is, if they really want to keep their state from drying up.

Acknowledgments

Florida's water issues are huge, complex, and controversial. Making my task easier in learning and reporting about them were the many and varied voices of Floridians who were willing to spend time with me and share their expertise and experience. Altogether I interviewed, consulted, or talked to more than one hundred people—hydrologists, biologists, engineers, water supply planners, water management district officials, boat captains, journalists, attorneys, riverkeepers, social scientists, elected officials, scuba divers, anglers, boaters, kayakers, and environmental activists. Over the past several years, I also attended numerous water forums, webinars, and public hearings where I heard scores of other water professionals and various stakeholders discuss Florida's water issues.

Thus, it would be long-winded on my part to mention everyone by name, though I offer all of them my gratitude for their help. However, there are some individuals who were overly generous with their time, patience, and assistance, and I would be remiss if I did not single them out.

It was Tallahassee water consultant Tom Singleton who opened my eyes to the importance of hydrology and how the relentless disruption and destruction of Florida's natural water processes lie at the core of our most pressing water problems. Special thanks also go to him for answering my weekly questions and critiquing an early draft of the book. University of South Florida hydrologists Shawn Landry and Mark Rains, along with environmental hydrologist Steve Leitman, also gave me much-valued insights into this expanding field of science.

To best grasp why Florida's springs are ailing, I turned repeatedly to Bob Knight, a scientist who has spent much of his career studying and defending them. I'm also indebted to former state springs biologist Jim Stevenson, a stern, experienced advocate for these beautiful yet stressed wonders, and to St. Johns River Water Management District springs director Casey Fitzgerald, for updating me about state-sponsored springs restoration efforts.

Many of Florida's underappreciated but dedicated Riverkeepers—such as Lisa Rinaman, Dan Tonsmeire, Rachel Silverstein, Laurie Murphy, John Cassani, and Neil Armingeon—supplied me with disturbing but welcomed insights about the status of many of Florida's abused rivers.

The never-ending arguments over water flow and algae outbreaks in Lake Okeechobee and the St. Lucie and Caloosahatchee waterways made much more sense to me after I interviewed Gary Goforth, an Everglades water resources engineer; Brian Lapointe, a research scientist with the Florida Atlantic University; Peter Barile, a senior scientist with Marine Research and Consulting in Melbourne; and Wendy Graham, the director of the University of Florida's Water Institute.

The issues facing Florida's farmers big and small deserve attention in any discussion about water, and I thank Charles Shinn, director of the Florida Farm Bureau Federation, and Rich Budell of the Florida Agricultural Council, for patiently explaining how growers view many of them.

During my research I discovered that social sciences now have much to say about water matters. Some of these insights found their way into my research, thanks to University of South Florida anthropologist Rebecca Zarger and University of Florida resource economist M. Jennison Kipp Searcy.

I now have a clearer picture of Florida's increasing use of recycled water after conversing with Melissa Meeker, former director of the South Florida Water Management District and now director of the WateReuse Association.

Former ranch owner Saundra Gray was kind enough to recall what it was like to participate at the St. Johns River Water Management District, while Margaret Spontak told me about working there as a water supply planner, and geologist Jim Gross contributed to my understanding by relating the tasks and challenges he encountered at the district as a scientist.

Janet Llewellyn, a recently retired DEP administrator, spent a lot of her time explaining to me what it is like to have the responsibility of dealing with Florida's water woes, and I am grateful to her.

Andrea Dutton, a University of Florida sea level rise expert, is both a researcher and an educator who boldly takes her climate change lessons beyond the classroom. I appreciate what she taught me during her public lectures and my interview with her. Tom Champeau, a Florida Fish and Wildlife Conservation Commission official, also provided me great assistance by explaining the unfolding impact of rising sea levels on Florida's flora and fauna, while Stuart Kennedy with the Miami Foundation revealed how optimistic policymakers are responding to the urban challenges resulting from climate change.

Seldom in the limelight is a galaxy of Floridians entrusted with the responsibility of keeping our freshwater, stormwater, and wastewater systems functioning every day. Several of them graciously made time for me, including Marion County's chief engineer Tracy Straub, Ocala Water and Sewer plant operator Chris Wilder, Ken Herd and Richard Menzies at Tampa Bay Water, and Cedar Key Water and Sewer District general manager John McPherson.

Any attempt to assess the state of Florida's waters demands mentioning ongoing water-related restoration projects. Thus, I'm grateful for briefings from Loisa Kerwin with the Florida Center for Environmental Studies at Florida Atlantic University about the Kissimmee Restoration Project, consultant Tony Janicki with the Tampa Bay Estuary Recover project, and Mark Sees, a key force behind the construction of the Orlando Wetlands Park.

Just about everyone I talked to for this book agreed the key to coming to grips with Florida's emerging water crisis is a political one. Helping me to better grasp how the art of politics is played in Tallahassee were State Representative Matt Caldwell and State Senators David Simmons and Linda Stewart. I'm thankful, too, to former Martin County Commissioner Anne Scott, and Heath Davis, mayor of Cedar Key, for their insights concerning water politics at the local level.

I also wish to recognize environmental attorneys David Guest, Heather Obara, Jerry Phillips, and Randie Denker for sharing their expertise concerning legal aspects of water pollution and water supply matters.

My gratitude is also extended to the staffers at the University Press of Florida (UPF) for their professional assistance, especially Director Meredith Babb for inviting me to submit a book proposal and then giving me a chance to write about water problems in my home state.

Several individuals read and critiqued my manuscript, and I am indebted to them for their informed opinions on how to make my book better. Among the UPF reviewers was Craig Pittman, one of the state's top environmental writers, whose editorial comments and reportage of the destruction of Florida's wetlands were indispensable to me. Copy editor Sharon Damoff deserves my thanks, too. My book is a much tighter, clearer, and more readable book than it would have been otherwise, because of her editing expertise. A salute also goes to UPF's Marthe Walters for fine-tuning the manuscript.

Another special reader was my wife, Susan Dunn, who uncomplainingly read my manuscript repeatedly and offered invaluable advice.

In the past year and a half, I read a small mountain's worth of printed and online material to prepare for writing this book. Among the most helpful of these sources were the works of fellow writer and friend Cynthia Barnett, whose book *Mirage* preceded my project by more than a decade. At the time, she eloquently and powerfully warned readers that water crises in Florida and the eastern United States were on the way. She was right.

My own interest in Florida water issues was nurtured long ago by two other friends. One is Ocalan Guy Marwick. Without his passionate activism, it is unlikely there would be a Silver River State Park or a Silver River Natural History Museum today. The other friend is Margy Bielling, a former University of Florida biology researcher, science teacher, and participant in the successful campaign to halt the ill-conceived Cross-Florida Barge Canal. Get the facts, she advised; don't compromise with the truth; and try to convince others with steady but polite persuasion.

I hope that *Drying Up* lives up to her expectation.

Selected Sources by Chapter

Introduction

Associated Press. "Crisis Feared as U.S. Water Supplies Dry Up." October 27, 2007. http://www.nbcnews.com/id/21494919/ns/us_news-environment/t/crisis-feared-us-water-supplies-dry/#.WIzxIoWcF9A.

Barnett, Cynthia. *Mirage: Florida and the Vanishing Water of the Eastern U.S.* Ann Arbor: University of Michigan Press, 2007.

"Florida Faces Vanishing Water Supply." National Public Radio, June 15, 2007. http://www.npr.org/templates/story/story.php?storyId=11097869.

Gillam, Carey. "Ogallala Aquifer: Could Critical Water Source Run Dry?" *Christian Science Monitor*, August 27, 2013. https://www.csmonitor.com/Environment/Latest-News-Wires/2013/0827/Ogallala-aquifer-Could-critical-water-source-run-dry.

McPherson, John. Telephone interview. 2017.

Chapter 1. Business as Usual?

Bystrak, Linda. E-mail correspondence. 2017.

"Deforestation Overview." World Wildlife Federation. https://www.worldwildlife.org/threats/deforestation.

Drake, Charles. AIF 2016 Water Forum Emerging Issues. https://www.youtube.com/watch?v=TdXjmdcwPTs.

Goforth, Gary. Personal, telephone, and e-mail correspondences. 2016–2018.

Hooke, Roger LeB., and Jose F. Martin-Duque. "Land Transformation by Humans: A Review." Geological Society of America. http://www.geosociety.org/gsatoday/archive/22/12/article/i1052-5173-22-12-4.htm.

"How Do Changes in Land Use Impact Water Resources?" American Geosciences Institute, National Academy of Sciences. https://www.americangeosciences.org/critical-issues/faq/how-do-changes-land-use-impact-water-resources.

"Improving Florida's Water Supply Management Structure: Ensuring and Sustaining Environmentally Sound Water Supplies and Resources to Meet Current and Future Needs." Florida Council of 100. http://www.fnai.fsu.edu/ARROW/Almanac/planning/waterreportfinal.pdf.

"International Decade for Action, Water for Life, 2005–2015: Water Scarcity." United Nations. http://www.un.org/waterforlifedecade/scarcity.shtml.

Katz, Vivian. Telephone and personal interview. 2016–2017.

Llewellyn, Janet. Telephone interview and e-mail correspondence. 2017–2018.

Monks, Kieron. "From Toilet to Tap: Getting a Taste for Drinking Recycled Waste Water." CNN, November 17, 2015. http://www.cnn.com/2014/05/01/world/from-toilet-to-tap-water/index.html.

"North Florida Regional Water Supply Plan." http://northfloridawater.com/watersupplyplan/index.html.

Pielou, E. C. *Fresh Water.* Chicago: University of Chicago Press, 1998.

Rains, Mark. Telephone interview. 2016.

Schers, Gerardus, Philip Waller, and Michael Condran. "The Future of Water Supply in Florida." https://www.wateronline.com/doc/the-future-of-water-supply-in-florida-0001.

Singleton, Tom. Telephone and e-mail interviews. 2016–2018.

Srinivasa, Veena, and Sharachchandra Lele. "Why We Must Have Water Budgets." *Hindu*, October 18, 2016. http://www.thehindu.com/opinion/columns/Why-we-must-have-water-budgets/article14179881.ece.

"Summary of Estimated Water Use in the United States in 2010." United States Geological Survey. http://pubs.usgs.gov/fs/2014/3109/pdf/fs2014-3109.pdf.

"Water 2070" (Webinar). 1000 Friends of Florida, February 15, 2016. https://handouts-live.s3.amazonaws.com/fdf2acce9c984c5fafa6e730e8e64f8d?sessionId=8563181653799509001&participantId=1000066.

"Water Crisis: Towards a Way to Improve the Situation." World Water Council. http://www.worldwatercouncil.org/library/archives/water-crisis.

"The World's Water." USGS Water Science School, United States Geological Survey. https://water.usgs.gov/edu/earthwherewater.html.

Chapter 2. Fouling the Waters?

"2016 Florida Infrastructure Report Card." American Society of Civil Engineers. http://www.infrastructurereportcard.org/florida/florida-infrastructure.

"Buried No Longer: Confronting America's Water Infrastructure Challenge." American Water Works Association. http://www.awwa.org/portals/0/files/legreg/documents/buriednolonger.pdf.

Burns, Eric. *The Spirits of America: A Social History of Alcohol.* Philadelphia: Temple University Press, 2004.

"Chemical Contaminants—HALs and Chemical Fact Sheets." Florida Department of Health. http://www.floridahealth.gov/search/search.cgi?zoom_query=edb&zcoom_cat%5B%5D=0.

Denker, Randie. Telephone interview. 2017.

"DEP's Pandering to Polluters, Says Former DEP Attorney." Florida Clean Water Network, January 9, 2017. http://floridacleanwaternetwork.org/dont-fall-for-deps-pandering-to-polluters-says-former-dep-attorney.

"Don't Drink the Water: Collapse of Florida's Safe Drinking Water Enforcement Program." August 2016. https://www.peer.org/assets/docs/fl/8_11_16_Dont_Drink_Water_report.pdf.

"Drinking Water & Human Health in Florida." Southern Regional Water Program. http://srwqis.tamu.edu/florida/program-information/florida-target-themes/drinking-water-and-human-health.

Duhigg, Charles. "Clean Water Laws Are Neglected, at a Cost in Suffering." *New York Times,* September 12, 2009. http://www.nytimes.com/2009/09/13/us/13water.html.

"Florida's Drinking Problem—Unsafe Water: Widespread Potable Water Violations but Virtually Nonexistent Enforcement." August 11, 2016. http://www.peer.org/news/news-releases/florida%E2%80%99s-drinking-problem-%E2%80%93-unsafe-water.html.

Johnson, Steve. *How We Got to Now: Six Innovations That Made the Modern World.* New York: Riverhead Books, 2014.

Lester, John. "Research Finds Alligator Problems Also Evident in Less Polluted Lakes." *Science News,* February 9, 1998. https://www.sciencedaily.com/releases/1998/02/980209154754.htm.

"List of Superfund Sites in Florida." U.S. Environmental Protection Agency. https://www.epa.gov/fl/list-superfund-sites-florida.

Llewellyn, Janet. Telephone and e-mail correspondences. 2017–2018.

"The Most Poorly Tested Chemicals in the World." Chemical Industry Archives. http://www.chemicalindustryarchives.org/factfiction/testing.asp.

Moran, John, "Water Decision Toxic for Democracy." *Tallahassee Democrat,* July 30, 2016. https://www.tallahassee.com/story/opinion/2016/07/30/water-decision-toxic-democracy/87748420/.

Murphy, Laurie. Telephone interview. 2016.

Obara, Heather. "Why Aren't Florida's Water Laws Protecting Florida's Water?" Water Voices Speakers Series, July 25, 2016. High Springs New Century Woman's Club. Ichetucknee Alliance.

Phillips, Jerry. Telephone and e-mail interviews. 2016–2017.

"Report Finds Water Pollution in Florida Costs up to $10.5 Billion, Annually." Earth Justice. November 27, 2012. http://earthjustice.org/news/press/2012/report-finds-water-pollution-in-florida-costs-up-to-10-5-billion-annually.

Spear, Kevin. "Famed for Alligator Research in Lake Apopka, Scientist Louis Guillette Dies." *Orlando Sentinel,* August 13, 2015. http://www.orlandosentinel.com/news/os-louis-guillette-scientist-dies-20150813-story.html.

Staletovich, Jenny. "Florida Drinking Water Ranks Among Nation's Worst, Study Finds." *Miami Herald,* May 2, 2017. http://www.miamiherald.com/news/local/environment/article148112799.html#storylink=cpy.

Stanton, Elizabeth A., and Matthew Taylor. "Valuing Florida's Clean Waters." Stockholm Environment Institute—U.S. Center. November 13, 2012. http://earthjustice.org/sites/default/files/ValuingFloridasCleanWaters.pdf.

Stewart, Linda. Telephone interview. 2017.

"Threats on Tap: Widespread Violations Highlight Need for Investment in Water Infrastructure and Protections." Natural Resources Defense Council, May 2017. https://www.nrdc.org/sites/default/files/threats-on-tap-water-infrastructure-protections-report.pdf.

"Threats to Florida's Water Quality, Natural Resources." UF/IFAS Extension Pinellas County. http://pinellas.ifas.ufl.edu/documents/ThreatstoFloridaWater.pdf.

"Toxic Waters." *New York Times*, May 22, 2012. https://www.nytimes.com/interactive/projects/toxic-waters/polluters/florida/index.html.

Waymer, Jim. "Florida Not Immune to Lead in Drinking Water." *Florida Today,* March 18, 2016. http://www.floridatoday.com/story/news/local/environment/2016/03/18/florida-not-immune-lead-drinking-water/81447772/.

Chapter 3. What's Nasty, Deadly, and Green All Over?

Doyle, Alistair. "Algae Can Evolve Rapidly to Cope with Climate Change, Study Finds." Huffington Post, November 15, 2014. http://www.huffingtonpost.com/2014/09/15/algae-evolve-climate-change_n_5818692.html.

"Drinking Water & Human Health in Florida." Southern Regional Water Program. http://srwqis.tamu.edu/florida/program-information/florida-target-themes/drinking-water-and-human-health.

"The Economic Benefits of Ecotourism." Florida Department of Economic Opportunity. http://www.floridajobs.org/community-planning-and-development/community-planning/community-planning-table-of-contents/ecotourism/the-economic-benefit-of-ecotourism.

Fitzgerald, Casey. Personal and e-mail correspondences. 2017.

"Florida's Springs: Protecting Nature's Gems." http://www.floridasprings.org/learn/journey/getting.

"Fritz Haber." *Science Heroes* website. http://scienceheroes.com.

Killer, Ed. "'The Blob' Gets Its 15 Minutes of Fame." *TCPalm*, July 2, 2016. http://archive.tcpalm.com/sports/columnists/ed-killer/ed-killer-the-blob-gets-its-15-minutes-of-fame-366cfcef-7609-02f0-e053-0100007f0e1b-385327211.html.

Knight, Robert L. Personal, telephone, and e-mail interviews. 2016–2017.

Knight, Robert L. *Silenced Springs: Moving from Tragedy to Hope.* Gainesville: Howard T. Odum Florida Springs Institute, 2015.

Knight, Robert L., and Ronald A. Clarke. "Florida Springs: A Water-Budget Approach to Estimating Water Availability." *Journal of Earth Science and Engineering* 6, no. 2 (February 2016). https://howardtodumfloridaspringsinstitute.wildapricot.org/resources/Pictures/Florida%20Springs%20Water%20Budget_JEASE-V01.6%20N0.22016.pdf.

McCasland, Margaret, Nancy M. Trautmann, Keith S. Porter, and Robert J. Wagenett. "Nitrate: Health Effects in Drinking Water." Cornell University, Cooperative Extension. http://psep.cce.cornell.edu/facts-slides-self/facts/nit-heef-grw85.aspx.

Miller, Dee Ann. E-mail correspondence. 2017.

"Movies and TV in Florida." Florida Memory, State Library and Archives of Florida. https://www.floridamemory.com/photographiccollection/photo_exhibits/movies.

"Nitrate/Nitrite-ToxFAQs." U.S. Centers for Disease Control. https://www.atsdr.cdc.gov/toxfaqs/tfacts204.pdf.

Rossiter, Fred. Telephone interview. 2017.

Staletovich, Jenny. "Keys Fishing Guides Send a Message to Tallahassee: Help!" *Miami Herald,* April 6, 2017. http://www.miamiherald.com/news/local/environment/article143226484.html.

Stevenson, Jim. Personal, telephone, and e-mail correspondences. 2017–2018.

Stevenson, Jim. "Recap of Springs Issues." Howard T. Odum Florida Springs Institute, March 8, 2013. https://www.youtube.com/watch?v=IOhWX5EnbGo&t=183s.

Chapter 4. Drain Me a River

Allman, T. D. *Finding Florida: The True History of the Sunshine State.* New York: Atlantic Monthly Press, 2013.

Buck, James. "Biscayne Sketches at the Far South" (1877). *Tequesta: The Journal of the Historical Association of Southern Florida* 29 (1979): 81. Digital Collection, Florida International University. http://digitalcollections.fiu.edu/tequesta/files/1979/79_1_07.pdf.

Grunwald, Michael. *The Swamp: The Everglades, Florida, and the Politics of Paradise.* New York: Simon and Schuster, 2006.

"Henry Morrison Flagler Biography." Henry Morrison Flagler Museum. https://flaglermuseum.us/history/flagler-biography.

Muir, John. *A Thousand-Mile Walk to the Gulf.* Boston: Houghton Mifflin, 1916.

Pittman, Craig, and Matthew Waite. *Paving Paradise: Florida's Vanishing Wetlands and the Failure of No Net Loss.* Gainesville: University Press of Florida, 2009.

Roberts, Diane. *Dream State: Eight Generations of Swamp Lawyers, Conquistadors, Confederate Daughters, Banana Republicans, and Other Florida Wildlife.* New York: Free Press, 2004.

Romm, Joe. "Wetlands Destruction—Another Climate Feedback." Think Progress. https://thinkprogress.org/wetlands-destruction-another-climate-feedback-5011ec695440.

Ruhl, J. B., and James Salzman. "The Effects of Wetland Mitigation Banking on People." Environmental Laws Institute, National Wetland Newsletter, March–April 2006. https://www.researchgate.net/publication/228322149_The_Effects_of_Wetland_Mitigation_Banking_on_People.

Simmons, David. Telephone interview. 2016.

Smith, Buckingham. "Reconnaissance of the Everglades, 1848." World Heritage Encyclopedia, Everglades Digital Library. Florida International University. http://dpanther.fiu.edu/sobek/PU00060001/00001.

Tebeau, Charlton W. *A History of Florida.* Coral Gables: University of Miami Press, 1971.

Chapter 5. Replumbing the Great Florida Outdoors

Antonini, Gustavo A., David A. Fann, and Paul Roat. *A Historical Geography of Southwest Florida Waterways.* Vol. 2, *Placida Harbor to Marco Island.* New York: Simon and Schuster, 1940. University of Florida Digital Collection, 2002. http://ufdc.ufl.edu/UF00093670/00002/pdf.

Carter, W. Hodding. *Stolen Water: Saving the Everglades from Its Friends, Foes, and Florida.* New York: Atria Books, 2004.

"The Cross Florida Barge Canal." Florida Frontiers, Florida Historical Society. https://myfloridahistory.org/frontiers/article/95.

Dahl, Thomas E., and Gregory J. Allord. "Technical Aspects of Wetlands: History of Wetlands in the Conterminous United States." National Water Summary on Wetland Resources, United States Geological Survey Water Supply Paper 2425. https://water.usgs.gov/nwsum/WSP2425/history.html.

"Florida Water Management." St. Johns River Water Management District. http://www.sjrwmd.com/history/1900-1949.html.

Kitchen, Sebastian. "Port Director: Channel Dredging of St. Johns River Could Begin This Year." *Florida Times-Union,* February 2, 2017. http://jacksonville.com/news/2017-02-01/port-director-channel-dredging-st-johns-river-could-begin-year.

"Lake Okeechobee and the Herbert Hoover Dike." U.S. Army Corps of Engineers, Jacksonville District. https://www.youtube.com/watch?v=KpkhJgV_mLo.

Landry, Clay J. "Who Drained the Everglades?" Property and Environment Research Center. https://www.perc.org/2002/03/01/who-drained-the-everglades.

Meltz, Robert. "Wetlands Regulation and the Law of Property Rights 'Takings.'" CRS Report for Congress, February 17, 2000. http://congressionalresearch.com/RL30423/document.php.

Pittman, Craig, and Matthew Waite. *Paving Paradise: Florida's Vanishing Wetlands and the Failure of No Net Loss.* Gainesville: University Press of Florida, 2009.

"Preserving Land Not Enough for Springs." *Ocala Star Banner,* June 22, 2017. https://www.ocala.com/opinion/20170622/editorial-preserving-land-not-enough-for-springs.

"Ruling Risks Retreat on Environmental Protection." *Tampa Bay Times,* June 28, 2013. http://www.tampabay.com/opinion/editorials/editorial-ruling-risks-retreat-on-environmental-protection/2129113.

"Status and Trends of Wetlands in the Coastal Watershed of the Conterminous United States, 2004 to 2009." National Oceanic and Atmospheric Administration National Marine Fisheries Service, and U.S. Department of the Interior, Fish and Wildlife Service. https://www.fws.gov/wetlands/Documents/Status-and-Trends-of-Wetlands-in-the-Coastal-Watersheds-of-the-Eastern-United-States-1998-to-2004.pdf.

"What Is a Wetland?" Environmental Protection Agency. https://www.epa.gov/wetlands/what-wetland.

Will, Lawrence E. *A Dredgeman of Cape Sable.* St. Petersburg, Florida: Great Outdoor Publishing, 1967.

Chapter 6. The Florida Growth Machine

"Central Florida's Reedy Creek Improvement District Has Wide-Ranging Authority." Office of Program Policy Analysis and Government Accountability, December 2004. http://www.oppaga.state.fl.us/reports/pdf/0481rpt.pdf.

Farago, Alan. "How the Growth Machine Ate Florida." *Counterpunch,* June 26, 2012. http://www.counterpunch.org/2012/06/26/how-the-growth-machine-ate-florida.

"Florida: Demographics." The Florida Legislature Office of Economic and Demographic Research, April 20–21, 2011. http://edr.state.fl.us/Content/presentations/population-demographics/DemographicOverview_4-20-11.pdf.

Harari, Yuval Noah. *Sapiens: A Brief History of Humankind.* New York: HarperCollins, 2015.

Hiassen, Carl. *Team Rodent: How Disney Devours the World.* New York: Ballantine, 1998.

"John DeGrove, Florida's Father of Growth Management, Dies." Naked Politics, *Miami Herald,* April 16, 2012. http://miamiherald.typepad.com/nakedpolitics/2012/04/john-degrove-floridas-father-of-growth-management-dies.html.

Koenig, David. *Realityland: True-Life Adventures at Walt Disney World.* Irvine, California: Bonaventure Press, 2007.

Lapidos, Juliet. "Why Is Florida God's Waiting Room?" *Slate,* January 29, 2008. http://www.slate.com/articles/news_and_politics/explainer/2008/01/why_is_florida_gods_waiting_room.html.

Mormino, Gary R. *Land of Sunshine, State of Dreams: A Social History of Modern Florida.* Gainesville: University Press of Florida, 2005.

Nolan, David. *Fifty Feet in Paradise: The Booming of Florida.* San Diego: Harcourt, Brace, Jovanovich, 1984.

Scott, Anne. Telephone interview. 2016.

Sweeney, Mark. Telephone interview. 2016.

Troxler, Howard. Public talk at the Hillsborough County League of Women Voters luncheon, August 10, 2010. https://www.youtube.com/watch?v=zjlapfSZ0qo.

Watkins, Tom. "Former Florida Gov. Reubin Askew Dies at 85." CNN, March 13, 2015. http://www.cnn.com/2014/03/13/politics/reubin-askew-death/index.html.

Wilson, Edward O. *The Meaning of Human Existence.* New York: Liveright, 2014.

Chapter 7. Growing the Water Pie

"Alternative Water Supplies." Florida Department of Environmental Protection, May 2014. https://www.dep.state.fl.us/water/waterpolicy/docs/factsheets/wrfss-alternative-water-supplies.pdf.

"Annual Status Report on Regional Supply Planning." Florida Department of Environmental Protection, December 2013. http://www.dep.state.fl.us/water/waterpolicy/docs/2013_annual_rwsp.pdf.

Aristotle. *Meteorologica.* Vol. VII. Translated by H.D.P. Lee. Loeb Classical Library 397. https://books.google.com/books/p/harvard?q=salt+water&vid=ISBN9780674994362&hl=en_US&ie=UTF-8&oe=UTF-8&btnG=Go.

Central Florida Water Initiative e-mail staff responses to author's questions. June 20, 2017.

Florida Geological Survey. Coastal Research Projects. http://dep.state.fl.us/geology/programs/coastal/coastal.htm.

"Florida Water Plan." Florida Department of Environmental Protection. https://floridadep.gov/water-policy/water-policy/content/florida-water-plan.

Gross, Jim. Telephone and e-mail interviews. 2017.

Herd, Ken. Personal interview. 2017.

"How Should Florida's Water Supply Be Managed in Response to Growth?" PURC/Askew Water Conference. March 31–April 1, 2005. https://warrington.ufl.edu/centers/purc/docs/PURC_AskewConference2005.pdf.

Kovski, Alan. "Florida Looks to Recycling, Desalination As It Faces Increasing Water Supply Crunch." *Bloomberg News*, January 29, 2015. https://www.bna.com/florida-looks-recycling-n17179922567.

Kumar, Manish, Tyler Culp, and Yuexiao Shen. "Water Desalination History, Advances, and Challenges." National Academy of Engineering. *The Bridge*, December 20, 2016. https://www.nae.edu/Publications/Bridge/164237/164313.aspx.

Lewis, Jamie, and Alan Wright. "Reclaimed Water Use for Edible Crop Production in Florida." University of Florida IFAS Extension. https://watereuse.org/wp-content/uploads/2015/01/IFAS-Reclaimed-Water-Use-for-Edible-Crop-Production-in-Florida.pdf.

Meeker, Melissa. Telephone and e-mail interview. 2017.

Menzies, Rich. Personal interview. 2017.

"Minimum Flows and Levels." Southwest Florida Water Management District. https://www.swfwmd.state.fl.us/projects/mfls.

Monson, Teresa. "St. Johns, Suwannee River Governing Boards Approve First-Ever Water Supply Plan for North Florida." St. Johns River Water Management District. http://webapub.sjrwmd.com/agws10/news_release/ViewNews.aspx?nrd=nr17-014.

"New Science or Same Ol' Politics on Sleepy Creek Permit?" *Ocala Star Banner*, January 8, 2017. http://www.ocala.com/opinion/20170108/editorial-new-science-or-same-ol-politics-on-sleepy-creek-permit.

"North Florida Regional Water Supply Plan." http://www.northfloridawater.com/watersupplyplan/index.html.

"North Florida Regional Water Supply Plan (2015–2035)." See review comments by Florida Springs Council. https://northfloridawater.com/watersupplyplan/documents/draft/DRAFT_NFRWSP_Appendices_01132017.pdf.

Rains, Mark. Telephone interview. 2016.

"Re-examining the Submarine Spring off Crescent Beach, Florida." U.S. Department of the Interior, U.S. Geological Survey, South Florida Information Access. https://sofia.usgs.gov/publications/ofr/00-158.

"Regional Water Supply Plan, 2015: A Comprehensive Plan for Orange, Osceola, Polk, Seminole, and Southern Lake Counties." Planning Document, Vol. 1. Central Florida Water Initiative. November 2015. http://cfwiwater.com/pdfs/plans/CFWI_RWSP_VolI_Final_2015-12-16.pdf.

"Regional Water Supply Planning, 2015 Annual Report." Florida Department of Environmental Protection. http://www.dep.state.fl.us/water/waterpolicy/docs/2015_Annual_Reg_Water_Supply.pdf.

"Regional Water Supply Planning, 2017 Annual Report." Florida Department of Environmental Protection. https://floridadep.gov/water-policy/water-policy/content/water-supply.

Register, Mike, Claire E. Muirhead, and Jason M. Mickel. Telephone interview. 2017.

Rimbey, Brad W. "Significant Harm." Swiftmud Springs Coast MFL Workshop. September 6, 2011. https://www.swfwmd.state.fl.us/media/1379.

Rogers, Paul. "Nation's Largest Ocean Desalination Plant Goes Up Near San Diego: Future of the California Coast?" *Mercury News*, January 23, 2017. http://www.mercurynews.com/2014/05/29/nations-largest-ocean-desalination-plant-goes-up-near-san-diego-future-of-the-california-coast.

Shinn, Charles. Telephone interview. 2016.

Spontak, Margaret. Personal and telephone interviews. 2017.

Straub, Tracy. Personal and e-mail interviews. 2016.

"Tampa Bay Seawater Desalination Plant." http://www.tampabaywater.org/documents/fact-sheets/desal-fact-sheet.pdf.

"Wakulla Spring." http://www.floridasprings.org/protecting/featured/wakulla.

Wildner, Chris. Personal interview. 2017.

Chapter 8. Whose Water Is It?

Blain, L. M. "Florida's Water: Who Owns It and By Whose Authority." *Water, Florida Law*. University of Florida Digital Collection. http://ufdc.ufl.edu/WL00000151/00001?search=owns+=water.

Bragg, Melvyn. "*Water.*" *In Our Time*, BBC Radio. http://www.bbc.co.uk/programmes/b01rgm9g.

Constitution of the State of Florida. https://www.flsenate.gov/Laws/Constitution#A2).

Hardin, Garrett. "The Tragedy of the Commons." December 13, 1968. The Garret Hardin Society. http://www.garretthardinsociety.org/articles/art_tragedy_of_the_commons.html.

"The Human Right to Water and Sanitation." General Assembly, United Nations, August 3, 2010. http://www.un.org/es/comun/docs/?symbol=A/RES/64/292&lang=E).

"Is Water a Human Right?" *On Earth*. Natural Resources Defense Council. https://www.nrdc.org/onearth/water-human-right.

Levin, Matt. "Rick Scott's Cuts Dilute Water Laws." *Broward Palm Beach New Times*, June 6, 2013. http://www.browardpalmbeach.com/news/rick-scotts-cuts-dilute-water-laws-6351355.

"Making Water a Human Right." United Nations Regional Information Centre for Western Europe. http://www.unric.org/en/water/27360-making-water-a-human-right.

Marshall, Skylar. "California Declares a Human Right to Water." *University of Denver Water Law Review*. Sturm College of Law. June 10, 2013. http://duwaterlawreview.com/ca-human-right-to-water.

Murthy, Sharmila L. "A New Constitutive Commitment to Water." *Boston College Journal of Law and Social Justice* 36 (2016): 159–233. http://lawdigitalcommons.bc.edu/jlsj/v0136/iss2/2.

Plager, Sheldon J., and Frank E. Maloney. "Florida's Streams—Water Rights in a Water Wonderland." Paper 1155, 1957. Articles by Maurer Faculty, Maurer School of Law, University of Indiana. http://www.repository.law.indiana.edu/facpub/1155.

"Progressive Realisation and Non-Regression." International Network for Economic, Social, and Cultural Rights. https://www.escr-net.org/resources/progressive-realisation-and-non-regression.

Prud'homme, Alex. *The Ripple Effect: The Fate of Freshwater in the Twenty-First Century.* New York: Scribner, 2011.

"The Right to Water, Fact Sheet No. 35." World Health Organization, United Nations. http://www.ohchr.org/documents/publications/factsheet35en.pdf.

Scott, Linda. Telephone interview. 2017.

Straub, Tracy. Personal interview. 2016.

"The Water in You." USGS Water Science School. https://water.usgs.gov/edu/propertyyou.html.

"Water Resources." 1000 Friends of Florida. http://www.1000friendsofflorida.org/saving-special-places/water-resources.

Chapter 9. Florida's Water Wars

Alvarez, Lizette. "A Fight Over Water, and to Save a Way of Life." *New York Times*, June 2, 2013. http://www.nytimes.com/2013/06/03/us/thirst-for-fresh-water-threatens-apalachicola-bay-fisheries.html.

"Chapter 373, F.S., Water Resources." *The Florida Senate Interim Report 2010-114*, September 2009. http://archive.flsenate.gov/data/Publications/2010/Senate/reports/interim_reports/pdf/2010-114ep.pdf, 9–10.

"Chattahoochee: Water War History." http://www.waterwar.org/history.html.

"Citizens: It'll Take a War to Get Our Water." *Suwannee Democrat and Jasper Mayo Free Press,* December 21, 2005. http://www.suwanneedemocrat.com/news/mayo_free_press/citizens-it-11-take-a-war-to-get-our-water/article_f9926bb3-e72f-5265-9117-7083fb163044.html.

Erskine, Mary Beth. "Preventing Water Wars." *University of South Florida News,* April 19, 2010. news.usf.edu/article/templates/?a=2239.

"Everglades Jetport." National Park Service. https://www.nps.gov/bicy/learn/historyculture/miami-jetport.htm.

"Global Water Security: Intelligence Community Assessment." February 2, 2012. https://www.dni.gov/files/documents/Special%20Report_ICA%20Global%20Water%20Security.pdf.

"Gov. Scott: Florida Will Take Historic Legal Action Against Georgia in Fight to Save Apalachicola." Florida Governor's Office, August 16, 2013. http://www.flgov.com/gov-scott-florida-will-take-historic-legal-action-against-georgia-in-fight-to-save-apalachicola-2.

Gross, Jim. "We Are Headed for Water Wars, Perhaps on an Unprecedented Scale." *Gainesville Sun,* April 7, 2017. http://www.gainesville.com/opinion/20170407/jim-gross-florida-headed-for-water-wars.

Hawthorne, Carnell. "North Florida Water Under Threat Again." *Suwannee Democrat and Jasper Mayo Free Press,* November 6, 2009. http://www.suwanneedemocrat.com/news/local_news/north-florida-water-under-threat-again/article_c877879f-5141-5484-b344-53f63aff0cd6.html.

"History of Big Cypress." http://www.evergladesonline.com/history-big-cypress.htm.

"Hydrologic Activity." Big Cypress, National Park Service. https://www.nps.gov/bicy/learn/nature/hydrologicactivity.htm.

"Improving Florida's Water Supply Management Structure: Ensuring and Sustaining Environmentally Sound Water Supplies and Resources to Meet Current and Future Needs." Florida Council of 100. http://www.fnai.fsu.edu/ARROW/Almanac/planning/waterreportfinal.pdf.

Leitman, Steve. Telephone interview. 2017.

Mitchell, John G. "The Bitter Struggle for a National Park." *American Heritage* 21, no. 3 (1970). http://www.americanheritage.com/content/bitter-struggle-national-park?page=2.

"Official Sides with Georgia in 'Water Wars.'" *Panama City News Herald,* February 14, 2017. http://www.newsherald.com/news/20170214/official-sides-with-georgia-in-water-wars.

"Our River." Chattahoochee Riverkeeper. https://chattahoochee.org/our-river.

Ritchie, Bruce. "Florida Asks U.S. Supreme Court to Save Apalachicola River, Oyster Industry." *Politico Florida,* June 1, 2017. http://www.politico.com/states/florida/story/2017/06/01/irreplaceable-apalachicola-river-at-risk-if-bad-recommendation-followed-florida-tells-us-supreme-court-112469.

"State of Florida, Plaintiff, v. State of Georgia, Defendant. Report of the Special Master." The U.S. Supreme Court. Ralph I. Lancaster Jr. February 14, 2017. www.scotusblog.com/wp-content/uploads/2017/09/2017.02.14-Report-of-Special-Master.pdf.

Tonsmeire, Dan. Telephone and e-mail interview. 2016.

Chapter 10. The Mother of All Florida Water Wars

"Another View: Toughen the Everglades Forever Act." *TCPalm,* January 5, 2017. http://www.tcpalm.com/story/opinion/editorials/2017/01/05/another-view-toughen-everglades-forever-act/96200710.

Brenner, Marie. "In the Kingdom of Big Sugar." *Vanity Fair,* February 2011. http://www.vanityfair.com/news/2001/02/floridas-fanjuls-200102.

Budell, Rich. Telephone interview. 2017.

Carter, W. Hodding. *Stolen Water: Saving the Everglades from Its Friends, Foes, and Florida.* New York: Atria Books, 2004.

Clement, Gail. "Reclaiming the Everglades: South Florida's Natural History, 1884 to 1934." Everglades Digital Library, Florida International University. http://everglades.fiu.edu/reclaim/timeline/timeline10.htm.

"Everglades Restoration: The Kissimmee River System." Florida Natural Resources Leadership Institute. http://nrli.ifas.ufl.edu/reports/NRLISebring08.pdf.

Frederick, Peter. "Restoring the Everglades Will Benefit Both Humans and Nature." *UF News,* June 3, 2016. http://news.ufl.edu/articles/2016/06/restoring-the-everglades-will-benefit-both-humans-and-nature.php.

Goforth, Gary. Personal, telephone, and e-mail correspondences. 2016–2018.

"Gov. Scott Declares State of Emergency in St. Lucie and Martin Counties Following Algal Blooms." Florida Governor's Office, News Release, June 29, 2016. http://www.flgov.com/2016/06/29/gov-scott-declares-state-of-emergency-in-st-lucie-and-martin-counties-following-algal-blooms.

Graham, Wendy. Telephone interview. 2017.

Green, Amy. "Southern Reservoir? Northern Reservoir? Research Says Florida Needs Both." WLRN Public Radio and Television, March 5, 2017. http://wlrn.org/post/southern-reservoir-northern-reservoir-research-says-florida-needs-both.

Guest, David. Telephone and personal interviews. 2016.

Klas, Mary Ellen. "Sugar's Decades-Long Hold Over Everglades Came with a Price." *Miami Herald,* July 11, 2016. http://www.miamiherald.com/news/local/environment/article88992067.html.

Lapointe, Brian. Personal, telephone, and e-mail correspondences. 2016–2018.

"Lawmakers in Florida House Shouldn't Squander Best Chance Yet to Help the Everglades." *Miami Herald,* April 22, 2017. http://www.miamiherald.com/opinion/editorials/article146109504.html.

MacKay, Buddy, with Rick Edmonds. *How Florida Happened.* Gainesville: University Press of Florida, 2010.

Markle, Whitey. Personal and telephone interviews. 2016–2018.

"Okeechobee Reservoir Bill Becomes Law." Florida Realtors website, May 10, 2017. http://www.floridarealtors.org/NewsAndEvents/article.cfm?id=351826.

Pearson, Daniel R. "U.S. Sugar Policy Is Not So Sweet." Cato Institute, February 14, 2015. https://www.cato.org/publications/commentary/us-sugar-policy-not-so-sweet.

Pittman, Craig. "Plan Re-emerges to Shift Water Resources Across Florida." *Tampa Bay Times,* October 30, 2009. https://doc.uments.com/d-plan-re-emerges-to-shift-water-resources-across-florida.pdf.

Smart, Gil. "What's Behind Push for Deep Injection Wells Near Lake O?" *TCPalm,* June 13, 2017. http://www.tcpalm.com/story/opinion/columnists/gil-smart/2017/06/13/gil-smart-whats-behind-push-deep-injection-wells-near-lake-o/389838001.

"South Florida Algae Crisis Special Report." WPTV News, West Palm Beach, July 18, 2016. http://www.flsenate.gov/Media/PressReleases/Show/2658https://www.youtube.com/watch?v=t6RKIDJlBsA.

"The South Florida Everglades Restoration Project." University of Texas at Austin, Civil, Architectural, and Environmental Engineering. http://www.ce.utexas.edu/prof/maidment/grad/dugger/GLADES/glades.html.

Staletovich, Jenny. "Lake Okeechobee: A Time Warp for Polluted Water." *Miami Herald,* August 13, 2016. http://www.miamiherald.com/news/local/environment/article95442427.html.

Staletovich, Jenny. "New Everglades Fix Calls for Flushing Water Deep Underground." *Miami Herald*, February 15, 2017. http://www.miamiherald.com/news/local/environment/article132963849.html.

Swihart, Tom. *Florida's Water: A Fragile Resource in a Vulnerable State.* New York: RFF Press, 2011.

"Toxic Water Crisis: A Florida Finale of Horror or Leadership? Where We Stand." *Orlando Sentinel*, July 9, 2017. http://www.orlandosentinel.com/opinion/os-ed-florida-water-disasters-20160717-story.html.

Treadway, Tyler. "Southern Reservoir Would Curb More Lake Okeechobee Discharge." *TCPalm*, February 24, 2017. http://www.tcpalm.com/story/news/local/indian-river-lagoon/health/2017/02/23/lake-okeechobee-discharges-corps-of-engineers/98002888.

Chapter 11. Flooding Time Again

"A Brief History of Climate Change." *BBC News*, September 2013. http://www.bbc.com/news/science-environment-15874560.

Cassani, John. Telephone interview and e-mail correspondence. 2016–2017.

Champeau, Tom. Telephone and e-mail interviews. 2016.

"Climate Change." 1000 Friends of Florida. http://www.1000friendsofflorida.org/?s=climate+change&x=0&y=0.

"Climate Change: 2014 Synthesis Report." Intergovernmental Panel on Climate Change. https://www.ipcc.ch/pdf/assessment-report/ar5/syr/AR5_SYR_FINAL_All_Topics.pdf 2.

"Climate Change and Florida." Environmental Protection Agency, September 1997. http://www.miamidade.gov/environment/library/brochures/climate-change-and-florida.pdf.

"Climate Change and Sea-Level Rise in Florida: An Update on the Effect of Climate Change on Florida's Ocean and Coastal Resources." Florida Oceans and Coastal Council, December 2010. http://www.dep.state.fl.us/oceanscouncil.

"Climate Change in South Florida." Florida Center for Environmental Studies. http://www.ces.fau.edu/climate_change.

"Climate Change, Water, and Risk: Current Water Demands Are Not Sustainable." Water Facts. National Resources Defense Council, July 2010. https://www.nrdc.org/sites/default/files/WaterRisk.pdf.

Davenport, Coral. "Miami Finds Itself Ankle-Deep in Climate Change Debate." *New York Times*, May 7, 2014. https://www.nytimes.com/2014/05/08/us/florida-finds-itself-in-the-eye-of-the-storm-on-climate-change.html?_r=0.

Dutton, Andrea. "Sea Level Rise in Florida: Where Are We Headed?" The Bob Graham Center, the University of Florida, August 30, 2016. http://www.bobgrahamcenter.ufl.edu/content/dr-andrea-dutton-sea-level-rise.

Dutton, Andrea. Telephone interview. 2017.

Florida Climate Institute website (a network of ten Florida universities). https://floridaclimateinstitute.org/resources/faqs#faqnoanchor.

"Florida's Bays and Estuaries Threatened by Global Warming: Fishing as We Know It Could Disappear." National Wildlife Federation, May 30, 2006. http://www.nwf.org/News-and-Magazines/Media-Center/News-by-Topic/Global-Warming/2006/05-30-06-Floridas-Bays-and-Estuaries-Threatened-by-Global-Warming.aspx.

"Fresh Water Is in Danger, Study Says." *Tampa Tribune*, December 17, 1988. University of Florida Digital Collection. http://ufdc.ufl.edu/WL00000960/00001.

Gillis, Chad. "Climate Change Will Force Florida's Local Governments to Act." News-Press.com, May 7, 2015. http://www.news-press.com/story/news/2015/05/07/climate-change-will-force-floridas-local-governments-act/70967032.

The Global Risks Report, 2016. 11th Edition. World Economic Forum. http://www3.weforum.org/docs/GRR/WEF_GRR16.pdf.

Goodell, Jeff. "Goodbye, Miami." *Rolling Stone*, June 20, 2013. http://www.rollingstone.com/politics/news/why-the-city-of-miami-is-doomed-to-drown-20130620.

Grosso, Richard. "Sea Level Rise Study for 1000 Friends." http://www.1000friendsofflorida.org/wp-content/uploads/2015/01/Grosso-sea-level.pdf.

Kennedy, Stuart. Telephone interview. 2017.

"NOAA Technical Report NOS CO-OPS 083: Global and Regional Sea Level Rise Scenarios for the United States." National Oceanic and Atmospheric Administration, January 2017. https://tidesandcurrents.noaa.gov/publications/techrpt83_Global_and_Regional_SLR_Scenarios_for_the_US_final.pdf.

"Our Changing Climate: The U.S. National Climate Assessment." NOAA, Climate Program Office. Earth Forum, January 2015. http://www.firstpreshc.org/wp-content/uploads/2015/01/NCA-Higgins-Earth-Forum-1-18-2015.pdf.

Rasmussen, Carol. "Study: Rising Seas Slowed by Increasing Water on Land." Jet Propulsion Laboratory and California Institute of Technology, February 11, 2016. https://www.jpl.nasa.gov/news/news.php?feature=5136.

Shanklin, Mary. "Flood Insurance Reforms Loom." *Orlando Sentinel,* July 25, 2017. http://www.orlandosentinel.com/classified/realestate/os-bz-flood-insurance-20170724-story.html.

Wanless, Harold. "Rising Sea Levels Will Be Too Much, Too Fast for Florida." *The Conversation*, May 28, 2014. http://theconversation.com/rising-sea-levels-will-be-too-much-too-fast-for-florida-27198.

Chapter 12. Rethinking Water

Britt, Mike. Telephone interview. 2016.

Brookes, Julian. "Why Water Is the New Oil." *Rolling Stone,* July 7, 2011. http://www.rollingstone.com/politics/news/why-water-is-the-new-oil-20110707.

Budell, Rich. Telephone interview. 2016.

Caldwell, Matt. Telephone interview. 2016.

Gleick, Peter. "The Soft Path for Water." Iwater, Salon Internacional Del Ciclo Integral Del Aqua. Barcelona, December 16, 2016. https://www.youtube.com/watch?v=jyvqXUjzeiE.

Landry, Shawn. Telephone interview. 2016.

Leahy, Stephen. *Your Water Footprint: The Shocking Facts About How Much Water We Use to Make Everyday Products.* Buffalo, New York: Firefly Books, 2014.

Mooney, Chris. "Want to Get Conservatives to Save Energy? Stop the Environmentalist Preaching." *Washington Post*, February 12, 2015. https://www.washingtonpost.com/news/energy-environment/wp/2015/02/12/the-best-way-to-get-conservatives-to-save-energy-is-to-stop-the-environmentalist-preaching/?utm_term=.a0725ae8932e.

O'Laughlin, Jay. Telephone interviews. 2016–2017.
"Preventing Water Wars." *USF News*, April 19, 2010. http://news.usf.edu/article/templates/?z=0&a=2239.
Prud'homme, Alex. *The Ripple Effect: The Fate of Freshwater in the Twenty-First Century*. New York: Scribner, 2011.
Searcy, Jennison Kipp. Telephone interview. 2016.
Singleton, Tom. Telephone and e-mail interviews. 2016–2018.
Wolff, Gary, and Peter H. Gleick. "The Soft Path for Water." In *The World's Waters, 2002–2003*. Pacific Institute. http://www.pacinst.org/app/uploads/2013/02/worlds_water_2002_chapter13.pdf.
World Water Council. "Water Crisis: Towards a Way to Improve the Situation." http://www.worldwatercouncil.org/library/archives/water-crisis.
Zarger, Rebecca. Telephone interview. 2016.

Chapter 13. All Is Not Lost

Bastian, Terry. Telephone interviews. 2016–2017.
"Ecosystem Restoration." U.S. Army Corps of Engineers, Jacksonville District. http://www.saj.usace.army.mil/Missions/Environmental/Ecosystem-Restoration.
Farr, James A., and O. Greg Brock. "Florida's Landmark Programs for Conservation and Recreation Land Acquisition." Florida Department of Environmental Protection. http://dep.state.fl.us/lands/AcqHistory.htm.
Janicki, Tony. Telephone interview. 2016.
Kerwin, Loisa. E-mail correspondence. 2017.
"Major Restoration Project Progress in 2016." South Florida Water Management District, January 5, 2017. http://myemail.constantcontact.com/SFWMD-Makes-Makes-Significant-Progress-on-Major-Restoration-Projects-in-2016.html?soid=1117910826311&aid=uhfXzwC5dp8.
Meindl, Christopher. "Water Growth and Tampa Bay." American Association for Geographers. February 19, 2014. http://news.aag.org/2014/02/water-growth-and-tampa-bay.
Sees, Mark. Telephone interview. 2016.
Spear, Kevin. "Kissimmee River Restoration." *Orlando Sentinel*, January 4, 2010. http://articles.orlandosentinel.com/2010-01-04/news/1001030075_1_ambitious-river-restoration-kissimmee-river-river-s-restoration.
"Springs Protection and Restoration Projects." Suwannee River Water Management District. http://www.mysuwanneeriver.com/index.aspx?nid=400.
"The Tampa Bay Estuary." University of South Florida. http://exhibits.lib.usf.edu/exhibits/show/ohp-tampabayestuary/holly-sue-greening/biography.
Wanielista, Martin. Telephone interview. 2017.
Waters, Hannah. "Bringing Back Tampa Bay's Seagrass." Ocean Portal, Smithsonian National Museum of Natural History. http://ocean.si.edu/ocean-news/bringing-back-tampa-bay%E2%80%99s-seagrass.

Chapter 14. Politics and Solutions

Allyn, Jan. Telephone interview. 2017.
Caldwell, Matt. Telephone interview. 2016.
Davis, Heath. Telephone interview. 2018.
Goforth, Gary. Personal, telephone, and e-mail correspondences. 2016–2018.
Gray, Saundra. Personal and telephone interviews. 2017.
Gross, Jim. Telephone interview and e-mail correspondences. 2017–2018.
Guest, David. Telephone interview. 2016.
Lapointe, Brian. Telephone interview and e-mail correspondences. 2016–2018.
Llewellyn, Janet. Telephone and e-mail correspondences. 2017–2018.
Rinaman, Lisa. Personal and telephone interviews. 2016–2017.
Singleton. Tom. Telephone interview and e-mail correspondences. 2016–2018.
Stevenson, Jim. Personal, telephone, and e-mail correspondences. 2017–2018.
Swihart, Tom. *Florida's Water: A Fragile Resource in a Vulnerable State.* New York: RFF Press, 2011.

Major Sources

Allan, Tony. *Virtual Water*. London: I. B. Tauris, 2011.

Allman, T. D. *Finding Florida: The True History of the Sunshine State*. New York: Atlantic Monthly Press, 2013.

"Alternative Water Supplies." Florida Department of Environmental Protection, May 2014. https://www.dep.state.fl.us/water/waterpolicy/docs/factsheets/wrfss-alternative-water-supplies.pdf.

"Annual Status Report on Regional Supply Planning," Florida Department of Environmental Protection, December 2013. http://www.dep.state.fl.us/water/waterpolicy/docs/2013_annual_rwsp.pdf.

Antonini, Gustavo A., David A. Fann, and Paul Roat. *A Historical Geography of Southwest Florida Waterways*. Vol. 2, *Placida Harbor to Marco Island*. New York: Simon and Schuster, 1940. University of Florida Digital Collection, 2002. http://ufdc.ufl.edu/UF00093670/00002/pdf.

Barlow, Maude. *Blue Covenant: The Global Water Crisis and the Coming Battle for the Right to Water*. New York: New Press, 2008.

Barnett, Cynthia. *Blue Revolution: Unmaking America's Water Crisis*. Boston: Beacon Press, 2011.

Barnett, Cynthia. *Mirage: Florida and the Vanishing Water of the Eastern U.S.* Ann Arbor: University of Michigan Press, 2007.

Blain, L. M. "Florida's Water: Who Owns It and By Whose Authority." *Water, Florida Law*. University of Florida Digital Collection. http://ufdc.ufl.edu/WL00000151/00001?search=owns+=water.

Buck, James. "Biscayne Sketches at the Far South" (1877). *Tequesta: The Journal of the Historical Association of Southern Florida* 29 (1979): 81. Digital Collection, Florida International University. http://digitalcollections.fiu.edu/tequesta/files/1979/79_1_07.pdf.

Burns, Eric. *The Spirits of America: A Social History of Alcohol*. Philadelphia: Temple University Press, 2004.

Carter, W. Hodding. *Stolen Water: Saving the Everglades from Its Friends, Foes, and Florida.* New York: Atria Books, 2004.

"Central Florida's Reedy Creek Improvement District Has Wide-Ranging Authority." Office of Program Policy Analysis and Government Accountability, December 2004. http://www.oppaga.state.fl.us/reports/pdf/0481rpt.pdf.

Christian-Smith, Juliet, and Peter H. Gleick. *A Twenty-First Century U.S. Water Policy.* New York: Oxford University Press, 2012.

"Climate Change." 1000 Friends of Florida. http://www.1000friendsofflorida.org/?s=climate+change&x=0&y=0.

"Climate Change: 2014 Synthesis Report." Intergovernmental Panel on Climate Change. https://www.ipcc.ch/pdf/assessment-report/ar5/syr/AR5_SYR_FINAL_All_Topics.pdf 2.

"Climate Change in South Florida." Florida Center for Environmental Studies. http://www.ces.fau.edu/climate_change.

"Climate Change, Water, and Risk: Current Water Demands Are Not Sustainable." Water Facts. National Resources Defense Council, July 2010. https://www.nrdc.org/sites/default/files/WaterRisk.pdf.

Dahl, Thomas E., and Gregory J. Allord. "Technical Aspects of Wetlands: History of Wetlands in the Conterminous United States." National Water Summary on Wetland Resources, United States Geological Survey Water Supply Paper 2425. https://water.usgs.gov/nwsum/WSP2425/history.html.

Douglas, Marjory Stoneman. *The Everglades: River of Grass.* Sarasota: Pineapple Press, 50th Anniversary Edition, 1997.

Dutton, Andrea. "Sea Level Rise in Florida: Where Are We Headed?" The Bob Graham Center, the University of Florida, August 30, 2016. http://www.bobgrahamcenter.ufl.edu/content/dr-andrea-dutton-sea-level-rise.

"Ecosystem Restoration." U.S. Army Corps of Engineers, Jacksonville District. http://www.saj.usace.army.mil/Missions/Environmental/Ecosystem-Restoration.

Farr, James A., and O. Greg Brock. "Florida's Landmark Programs for Conservation and Recreation Land Acquisition." Florida Department of Environmental Protection. http://dep.state.fl.us/lands/AcqHistory.htm.

"Florida Water Plan." Florida Department of Environmental Protection. https://floridadep.gov/water-policy/water-policy/content/florida-water-plan.

"Global and Regional Sea Level Rise Scenarios for the United States." National Oceanic and Atmospheric Administration, January 2017. https://tidesandcurrents.noaa.gov/publications/techrpt83_Global_and_Regional_SLR_Scenarios_for_the_US_final.pdf.

The Global Risks Report, 2016. 11th Edition. World Economic Forum. http://www3.weforum.org/docs/GRR/WEF_GRR16.pdf.

"Global Water Security: Intelligence Community Assessment." February 2, 2012. https://www.dni.gov/files/documents/Special%20Report_ICA%20Global%20Water%20Security.pdf.

Godfrey, Matthew C., and Theodore Catton. *River of Interest: Water Management in*

South Florida and the Everglades, 1948–1950. Jacksonville: Government Printing Office for the U.S. Army Corps of Engineers, 2011.

Goodell, Jeff. *The Water Will Come: Rising Seas, Sinking Cities, and the Remaking of the Civilized World.* New York: Little, Brown and Company, 2017.

Grunwald, Michael. *The Swamp: The Everglades, Florida, and the Politics of Paradise.* New York: Simon and Schuster, 2006.

Harari, Yuval Noah. *Sapiens: A Brief History of Humankind.* New York: HarperCollins, 2015.

Hiassen, Carl. *Team Rodent: How Disney Devours the World.* New York: Ballantine, 1998.

Hine, Albert C., Donald P. Chambers, Tonya D. Clayton, Mark R. Hafen, and Gary T. Mitchum. *Sea Level Rise in Florida: Science, Impacts, and Options.* Gainesville: University Press of Florida, 2016.

"Hurricane of 1928." Historical Society of Palm Beach. http://www.pbchistoryonline.org/page/agriculture.

Hurston, Zora Neale. *Their Eyes Were Watching God.* New York: HarperPerennial, 2006.

"Improving Florida's Water Supply Management Structure: Ensuring and Sustaining Environmentally Sound Water Supplies and Resources to Meet Current and Future Needs." Florida Council of 100. http://www.fnai.fsu.edu/ARROW/Almanac/planning/waterreportfinal.pdf.

Johnson, Steve. *How We Got to Now: Six Innovations That Made the Modern World.* New York: Riverhead Books, 2014.

Knight, Robert L. *Silenced Springs: Moving from Tragedy to Hope.* Gainesville: Howard T. Odum Florida Springs Institute, 2015.

Knight, Robert L., and Ronald A. Clarke. "Florida Springs: A Water-Budget Approach to Estimating Water Availability." *Journal of Earth Science and Engineering* 6, no. 2 (February 2016): 59.

Koenig, David. *Realityland: True-Life Adventures at Walt Disney World.* Irvine, California: Bonaventure Press, 2007.

Lanz, Klaus. *The Greenpeace Book of Water.* New York: Sterling, 1995.

Leahy, Stephen. *Your Water Footprint: The Shocking Facts About How Much Water We Use to Make Everyday Products.* Buffalo, New York: Firefly Books, 2014.

MacKay, Buddy, with Rich Edmonds. *How Florida Happened.* Gainesville: University Press of Florida, 2010.

"Major Restoration Project Progress in 2016." South Florida Water Management District, January 5, 2017. http://myemail.constantcontact.com/SFWMD-Makes-Makes-Significant-Progress-on-Major-Restoration-Projects-in-2016.html?soid=1117910826311&aid=uhfXzwC5dp8.

Mitchell, John G. "The Bitter Struggle for a National Park." *American Heritage* 21, no. 3 (1970). http://www.americanheritage.com/content/bitter-struggle-national-park?page=2.

Mormino, Gary R. *Land of Sunshine, State of Dreams: A Social History of Modern Florida.* Gainesville: University Press of Florida, 2005.

Muir, John. *A Thousand-Mile Walk to the Gulf.* Boston: Houghton Mifflin, 1916.

Murthy, Sharmila L. "A New Constitutive Commitment to Water." *Boston College Journal of Law and Social Justice* 36 (2016): 159–233. http://lawdigitalcommons.bc.edu/jlsj/v0136/iss2/2.

Nixon, Richard. "Statement about Halting Construction of the Cross Florida Barge Canal." January 19, 1971. Online by Gerhard Peters and John T. Woolley, *The American Presidency Project.* http://www.presidency.ucsb.edu/ws/?pid=3044.

Nolan, David. *Fifty Feet in Paradise: The Booming of Florida.* San Diego: Harcourt, Brace, Jovanovich, 1984.

Noll, Steven, and David Tegeder. *Ditch of Dreams: The Cross Florida Barge Canal and the Struggle for Florida's Future.* Gainesville: University Press of Florida, 2009.

"North Florida Regional Water Supply Plan." http://northfloridawater.com/watersupplyplan/index.html.

"North Florida Regional Water Supply Plan (2015–2035)." See review comments by Florida Springs Council. https://northfloridawater.com/watersupplyplan/documents/draft/DRAFT_NFRWSP_Appendices_01132017.pdf.

"Our Changing Climate: The U.S. National Climate Assessment." NOAA, Climate Program Office. Earth Forum, January 2015. http://www.firstpreshc.org/wp-content/uploads/2015/01/NCA-Higgins-Earth-Forum-1-18-2015.pdf.

Pielou, E. C. *Fresh Water.* Chicago: University of Chicago Press, 1998.

Pittman, Craig, and Matthew Waite. *Paving Paradise: Florida's Vanishing Wetlands and the Failure of No Net Loss.* Gainesville: University Press of Florida, 2009.

Plager, Sheldon J., and Frank E. Maloney. "Florida's Streams—Water Rights in a Water Wonderland." Paper 1155, 1957. Articles by Maurer Faculty, Maurer School of Law, University of Indiana. http://www.repository.law.indiana.edu/facpub/1155.

Prud'homme, Alex. *The Ripple Effect: The Fate of Freshwater in the Twenty-First Century.* New York: Scribner, 2011.

Purdum, Elizabeth D. *Florida Waters: A Water Resources Manual from Florida's Water Management Districts.* Tallahassee: Florida State University, 2002.

"Regional Water Supply Plan, 2015: A Comprehensive Plan for Orange, Osceola, Polk, Seminole, and Southern Lake Counties." Planning Document, Vol. 1. Central Florida Water Initiative. November 2015. http://cfwiwater.com/pdfs/plans/CFWI_RWSP_VolI_Final_2015-12-16.pdf.

"Regional Water Supply Planning, 2015 Annual Report." Florida Department of Environmental Protection. http://www.dep.state.fl.us/water/waterpolicy/docs/2015_Annual_Reg_Water_Supply.pdf.

"Regional Water Supply Planning, 2017 Annual Report." Florida Department of Environmental Protection. https://floridadep.gov/water-policy/water-policy/content/water-supply.

Roberts, Diane. *Dream State: Eight Generations of Swamp Lawyers, Conquistadors, Confederate Daughters, Banana Republicans, and Other Florida Wildlife.* New York: Free Press, 2004.

"The South Florida Everglades Restoration Project." University of Texas at Austin, Civil, Architectural, and Environmental Engineering. http://www.ce.utexas.edu/prof/maidment/grad/dugger/GLADES/glades.html.

"Springs Protection and Restoration Projects." Suwannee River Water Management District. http://www.mysuwanneeriver.com/index.aspx?nid=400.

Stanton, Elizabeth A., and Matthew Taylor. "Valuing Florida's Clean Waters." Stockholm Environment Institute—U.S. Center. November 13, 2012. http://earthjustice.org/sites/default/files/ValuingFloridasCleanWaters.pdf.

"State of Florida, Plaintiff, v. State of Georgia, Defendant. Report of the Special Master." The U.S. Supreme Court. Ralph I. Lancaster Jr. February 14, 2017. www.scotusblog.com/wp-content/uploads/2017/09/2017.02.14-Report-of-Special-Master.pdf.

"Status and Trends of Wetlands in the Coastal Watershed of the Conterminous United States, 2004 to 2009." National Oceanic and Atmospheric Administration National Marine Fisheries Service, and U.S. Department of the Interior, Fish and Wildlife Service. https://www.fws.gov/wetlands/Documents/Status-and-Trends-of-Wetlands-in-the-Coastal-Watersheds-of-the-Eastern-United-States-1998-to-2004.pdf.

Stevenson, Jim A. *My Journey in Florida's State Parks: A Naturalist's Memoir.* San Bernardino, California: self-published, 2014.

Swanson, Peter. *Water: The Drop of Life.* Minnetonka, Minnesota: NorthWord Press, 2001.

Swihart, Tom. *Florida's Water: A Fragile Resource in a Vulnerable State.* New York: RFF Press, 2011.

"The Tampa Bay Estuary." University of South Florida. http://exhibits.lib.usf.edu/exhibits/show/ohp-tampabayestuary/holly-sue-greening/biography.

Tebeau, Charlton W. *A History of Florida.* Coral Gables: University of Miami Press, 1971.

"Water 2070" (Webinar). 1000 Friends of Florida, February 15, 2016. https://handouts-live.s3.amazonaws.com/fdf2acce9c984c5fafa6e730e8e64f8d?sessionId=8563181653799509001&participantId=1000066.

Whitney, Ellie, D. Bruce Means, and Anne Rudloe. *Florida's Waters.* Vol. 3 of *Florida's Natural Ecosystems and Native Species.* Sarasota: Pineapple Press, 2014.

Will, Lawrence E. *A Dredgeman of Cape Sable.* St. Petersburg, Florida: Great Outdoor Publishing, 1967.

Wilson, Edward O. *The Meaning of Human Existence.* New York: Liveright, 2014.

Wolff, Gary, and Peter H. Gleick. "The Soft Path for Water." In *The World's Waters, 2002–2003.* Pacific Institute. http://www.pacinst.org/app/uploads/2013/02/worlds_water_2002_chapter13.pdf.

Photo Credits

Page 121: By permission of the Southwest Florida Water Management District.

Page 122, *top*: By permission of the Earth Science and Remote Sensing Unit, NASA Johnson Space Center.

Page 122, *bottom*: Photo and permission by Jim Stevenson.

Page 123: State Archives of Florida, Florida Memory, https://www.floridamemory.com/items/show/31121.

Page 124: Photo by Dr. David E. LaHart, State Archives of Florida, Florida Memory, https://www.floridamemory.com/items/show/96427.

Page 125, *top*: Photo courtesy of Karen Chadwick.

Page 125, *bottom*: Photo and permission by Guy H. Means, Florida Geological Survey.

Page 126: Photo by Ted Lagerberg, State Archives of Florida, Florida Memory, https://www.floridamemory.com/items/show/82754.

Page 127, *top*: Photo by author.

Page 127, *bottom*: By permission of the Northwest Florida Water Management District.

Page 128, *top*: Photo by author.

Page 128, *bottom*: Photo and permission by Guy H. Means, Florida Geological Survey.

Page 129, *top*: Photo by Sandy Gandy, State Archives of Florida, Florida Memory, https://www.floridamemory.com/items/show/26553.

Page 129, *bottom*: State Archives of Florida, Florida Memory, https://www.floridamemory.com/items/show/335273.

Page 130, *top*: Photo by author.

Page 130, *bottom*: State Archives of Florida, Florida Memory, https://www.floridamemory.com/items/show/1276.

Page 131, *top*: State Archives of Florida, Florida Memory, https://www.floridamemory.com/items/show/255401.

Page 131, *bottom*: By W. A. Fishbaugh, State Archives of Florida, Florida Memory, https://www.floridamemory.com/items/show/30639.

Page 132, *top*: State Archives of Florida, Florida Memory, https://www.floridamemory.com/items/show/35064.

Page 132, *bottom*: Photo by Donn Dughi, State Archives of Florida, Florida Memory, https://www.floridamemory.com/items/show/103483.

Page 133, *top*: Photo and permission by Guy H. Means, Florida Geological Survey.

Page 133, *bottom*: By permission of Tampa Bay Water.

Page 134, *top*: By permission of Tampa Bay Water.

Page 134, *bottom*: By permission of the St. Johns River Water Management District.

Page 135, *top*: State Archives of Florida, Florida Memory, https://www.floridamemory.com/items/show/129059.

Page 135, *bottom*: By permission of the St. Johns River Water Management District.

Page 136, *top*: Photo by Bill Cotterell, State Archives of Florida, Florida Memory, https://www.floridamemory.com/items/show/269273.

Page 136, *bottom*: Photo by author.

Page 137, *top*: Shutterstock.

Page 137, *bottom*: Photo and permission by Guy H. Means, Florida Geological Survey.

Page 138: State Archives of Florida, Florida Memory, https://www.floridamemory.com/items/show/5621.

Page 139: By permission of the National Park Service.

Page 140, *top*: Photo by Bill Perry, by permission of the National Park Service.

Page 140, *bottom*: Photo by Rodney Cammauf, by permission of the National Park Service.

Page 141, *top*: Photo by Rodney Cammauf, by permission of the National Park Service.

Page 141, *bottom*: Shutterstock.

Page 142, *top*: Photo by author.

Page 142, *bottom*: By permission of the South Florida Water Management District.

Page 143: Photos and permission by Mike Britt.

Page 144: By permission of the National Park Service.

Page 145: By permission of the National Park Service.

Page 146, *top*: By permission of the South Florida Water Management District.

Page 146, *bottom*: By permission of the South Florida Water Management District.

Page 147, *top*: By permission of the City of Orlando.

Page 147, *bottom*: By permission of the City of Orlando.

Page 148, *top*: Photo by Brian Cousin. By permission of Harbor Branch Oceanographic Institute, Florida Atlantic University.

Page 148, *bottom*: Photo by Lovett E. Williams, State Archives of Florida, Florida Memory, https://www.floridamemory.com/items/show/96110.

Page 149: Photo and permission by Bridget Dunn.

Index

Page numbers in *italics* refer to photos.

JOHN M. DUNN has written more than four hundred articles for periodicals such as *Europe, Overseas Life, Sierra, Florida Trend, Business Florida,* and the *St. Petersburg Times*. He has also authored seventeen nonfiction young adult books, including *The Russian Revolution, The Relocation of the North American Indian, The Spread of Islam, The Civil Rights Movement,* and *José Martí: Cuba's Greatest Hero.* Several of his books have received national recognition. He lives with his wife in Ocala, Florida.